The
Theory of Total Consonance

The Theory
of Total Consonance

PAUL ROSBERGER

Rutherford • Madison • Teaneck
Fairleigh Dickinson University Press

© 1970 by Associated University Presses, Inc.

Library of Congress Catalogue Card Number: 71-92560

Associated University Presses, Inc.
Cranbury, New Jersey 08512

SBN: 8386 7570 0

Printed in the United States of America

To the memory of my aunt, Ruth Fisher

Contents

Contents

The
Theory of Total Consonance

1
Preliminary Background

Music is a source of emotional inspiration. For most of us it is a natural part of our education, inexplicably linked to the composer's ability in combining frequencies to which we react. Nor can this mysterious art rely on customary responses, for what is most often great is most often unique. And what is great is almost unanimously recognized without question. Yet we must admit that as in any fine art, but especially in music, its origins extend from science.

Music is not a pure science. You cannot apply scientific principles to music and automatically produce an improvement. A theoretical abstraction finds little sympathy with the ear. The ear was here before the art, so we fashion the art primarily to suit the ear. Initially this basis involves what the school of Pythagoras observed long ago: the very foundation of music, a relation of frequencies, must appear combined in simple, or low, ratios in order to provide maximum concord. And what is most pleasant is most resonant.

Two frequencies are resonant when they are in simple ratio with one another and their combination sounds consonant. The opposite of consonance is dissonance, caused either by the relation of very close fundamental frequencies or by their closely related harmonics. The closely related harmonics can produce either beats or new tones of a high order. Let us initially cover conditions that oppose consonant relations before proceeding to how consonant systems have been produced according to Pythagoras' postulate.

The most elemental frequency is a sine wave. If two sine waves of the same frequency are combined and are in phase with each other, a perfect concord is produced. Graphically, given that both are of the same amplitude, they will superimpose. If the frequency of one is fixed and the other is now raised slightly, beats will occur whose rate will be the difference between the two fundamental frequencies. Because the two frequencies are now out of phase with each other in a continuously varying degree, they will alternately tend to cancel and reinforce each other. Each period of varying amplitude is heard as a beat. The varying phase shift is modulating the amplitude of the combined frequencies.

Beats such as those just mentioned occur in music. But a musical tone is rarely composed of a single sine wave. Included with the fundamental frequency are harmonics. Harmonics are additional sine waves occurring simultaneously at frequencies related to the fundamental. Essentially, the harmonics present and their individual amplitudes determine the overall wave shape. These variables in turn give an identity, or timbre, to the tone generator. The potential variations are infinite.

Harmonics are an important, integral part of the sounds used in music. They are an especially important consideration in the combination of tones. Tones can be combined in simple ratio to produce a pair of notes called a diad. The diad will become less consonant as the simple ratio governing the fundamentals of its tones is increased. The reason for this dissonance can be adequately illustrated by comparison to the first six harmonics of the two tones. Further extension of harmonics to be compared will produce more relations. However, the relative level of consonance at which a pair of tones combine is quite accurately indicated without considering all of the possible harmonic interrelations of the diad.

After the unison, the next smallest ratio and most consonant diad is the octave, which is 2/1, or,

$$\frac{2f_x}{1}$$

where f_x is the lower frequency. The ratios continue as follows:

f_x	$2f_x$	$3f_x$	$4f_x$	$5f_x$	$6f_x$
	$2f_x$		$4f_x$		$6f_x$

All the harmonics shown correspond to one another and therefore will not cause beats. By reinforcing themselves they tend to reinforce the fundamentals because the frequency spacing between two adjacent harmonics of a tone *is* the fundamental frequency. Even if the fundamental frequency were to be removed, the consistent frequency spacing of the harmonics would create the impression in the ear that the fundamental was present. Such phenomena occur, for example, in hearing telephone reproduction and bells.

Following the octave in greatest consonance is the fifth, which is 3/2, or

$$\frac{3f_x}{2}$$

as follows:

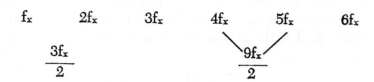

The relation of $4f_x$ and

$$\frac{9f_x}{2}$$

can beat and in the first order would result as the difference

$$\frac{f_x}{2}$$

and as the sum

$$\frac{17f_x}{2}$$

Also beating can occur with

$$\frac{9f_x}{2}$$

and $5f_x$. Some new frequencies have now been created in this process which are not the same as any of those initially generated and which therefore become potential sources of dissonance.

With the fourth, which is 4/3, or

$$\frac{4f_x}{3}$$

we have these relations:

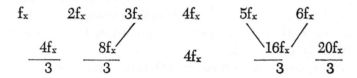

Already with only three examples, the possibilities for beating have grown tremendously, and more important, new frequencies have been generated not having equivalents in either the fundamental or harmonics of the original two tones. It is the increase in dissonance that reduces the concord of diads as the simple ratio is increased. This interrelation holds true for chords as well but on a correspondingly more complex level. It is these included intermodulation dissonances, as well as the degree of intermodulation controlled by what harmonics are present and the amplitudes, that give music its color. The merit of a composer can in one way be measured by his ability to combine and properly balance the different wave shapes of various instruments simultaneously. This is music.

But what about the beating that occurs if the fundamentals are not quite related in simple ratios? Paradoxically, the lower the ratio of the diad the greater the degree of "roughness" that appears when one of the notes is altered slightly in frequency. That is, the greatest consonance is bounded by the greatest dissonance. This emphasizes the importance of keeping frequencies exactly in the simple ratio relations that they were intended to have. A diad or chord favoring a consonant relationship of notes can easily result in a rough or beating sound. Later on it will be shown how systems of temperament intentionally shift these simple ratios into this area of "roughness." It would seem bad enough that this roughness can occur by the varying effects of environment that cause conventional instruments to go out of tune without deliberately creating it.

Starting with the octave, or 2/1 ratio, and referring to the previous example, the harmonics correspond fully, that is, $2f_x$, $4f_x$, $6f_x$, and so reinforce the fundamental. However, should one of the fundamentals be even slightly changed in frequency, the large quantity of harmonics of the two tones that previously were an exact match now become a large quantity of different frequencies in very close relations. The result is severe beating and discord. Where the exact octave relation of the two tones shown has three points of coincidence, the fifth shown has only two, that is, $3f_x$, $6f_x$. The exact fifth is less consonant than the exact octave. In the example, a slight frequency deviation will result in beating at the two closely related harmonics. This is one less harmonic parallel than the octave and therefore the fifth, although less consonant than the octave, will also have fewer sources of potential dissonance. The beat rate of a mistuned diad is also an important factor. For example, the intervals of an octave in equal temperament will each have different beat rates, and some rates are more noticeable than others.

Although each interval has its own different combinations that can cause beats, a specific example of beating in the 3/2 simple ratio is provided here to illustrate the mechanics of the general problem. Frequency is in cycles/second.

Given a concord:

100	200	300	. . .
	150	300	. . .

The *higher* frequency raised 1 c.p.s.:

100	200	300	. . .
	151	302	. . .

The first harmonic of the higher frequency now beats at 2 c.p.s. with the second harmonic of the lower frequency.

The *lower* frequency is raised 1 c.p.s.:

101	202	303	. . .
	150	300	. . .

The first harmonic again beats with the second harmonic but in this case the beat frequency is 3 c.p.s.

In music, the diad that provides the maximum concord is the unison, or one to one relation, followed by the octave and its doublings, which are, 2/1, 4/1, 8/1, etc. However, these consonant relations offer little variety of sound, which led Pythagoras (c. 582-509 B.C.) to assemble a scale that would provide notes in between the frequency span of the octave. These additional intervals can then be repeated in their respective positions throughout adjacent octaves. For this purpose the next most consonant interval after the octave was utilized, the fifth, or 3/2. By a variety of ascending fifths and descending octaves or descending fifths and ascending octaves a scale was constructed within one octave. Using for example the C major scale as it is currently termed, in order to locate D above C would start at C as 3/2 x 3/2 x 1/2, or, two fifths up and an octave down. This would make D 9/8 of C. C to D is therefore not as resonant as the C to G because the ratio of 9/8 is a higher ratio than 3/2 but is not an unreasonably high ratio. By the same process if we derive A above C we ascend from C by 3/2 to G by 3/2 to D by 3/2 to A by 1/2 to A an octave down. This makes A 27/16. This is not so simple a ratio. One more example, deriving B above C is

$$\frac{\left(\dfrac{3}{2}\right)^5}{\left(\dfrac{2}{1}\right)^2}$$

or 243/128. Although originally this system worked well enough for musical needs, Pythagoras had not actually followed through with his goal of intervals in simple ratio. It was to be a long time before the advent of harmony when the deficiencies of these high ratios would need correction.

A couple of millenia after Pythagoras, Zarlino (Maître de Chapelle at St. Mark's, Venice, ca. 1560) modified the Pythagorean scale primarily to reduce some of the higher ratios to more resonant, lower ones. He did this by rounding, for example C to A, 27/16 became 5/3. The difference can be compared with a common denominator as 27/16 equals 81/48, 5/3 equals 80/48 or, 1 part in 48. Likewise, C to E changed from 81/64 to 5/4, and C to B from 243/128 to 15/8.

To this proceeding scale more steps were added to arrive at the current twelve. For example, a major sixth below the fundamental frequency of C is E flat, which is then raised an octave. That is, a major sixth down, which is the inverted ratio of a major sixth up, or 3/5, times an octave up, or 2/1, equals 6/5. However, more than twelve steps would be in use today if some form of temperament had not been utilized. A temperament of frequencies extends from the need to modulate. Given a scale of frequencies, an important means of avoiding the redundant outlining of only one scale in the composition comes by raising or lowering all the scale steps by an equal amount of frequency. Relations by ratio to one another will remain the same but the individual frequencies have been changed. However, modulating frequencies is not an easy task when we are dealing with instruments having means of frequency generation that are fixed for each individual frequency. While this is a shortcoming, nevertheless, fixed frequency instruments form a very large part of the instrumental ensemble and are commonly used by the composer for working out compositions. In their favor, such instruments are capable of providing an *exact* simple frequency ratio if so constructed. Such accuracy can be obtained with ease in comparison to locating exact ratios on variable pitch instruments. For that matter, performers of variable pitch instruments can only approach an exact simple ratio; their degree of success is a measure of their experience.

It is logical with fixed pitch instruments to modulate

by steps that are already present. For example, if a hypothetical scale were constructed of fixed frequencies f_1, f_2, f_3, where f_3 is smaller than $2f_1$, then an example of modulation would be f_2, f_3, $2f_1$. Unfortunately, correct modulation can not come about that easily. To be musically correct intervals of the scale need be only related to the fundamental frequency in simple ratios. To be able to modulate, they must now be compared to one another. By adding five additional notes, termed the accidentals, where there were wide gaps in the scale of seven notes, the new chromatic scale produced still does not have equal frequency separation from frequency to frequency. Further, the original scale of seven notes was never intended to have equal frequency separation from step to step. Therefore, an attempt to modulate the original scale by starting it on any one of the eleven remaining notes is unsuccessful. The frequency spacings will not superimpose on one another. Also, in providing steps for a scale that has been modulated from the original frequencies appear that are close to one another but not the same. These deviations have been smoothed over in our present use of a temperament by choosing a common frequency for, say, both a C sharp and a D flat.

Wrestling out of the emphasis on modulation but still retaining a reasonable number of subdivisions within the octave was the solution of a temperament. A temperament adjusts the true frequency ratios, some or all, depending on the temperament. The adjustment is made under the premise that errors can be acceptably introduced up to the limit that the "temperament" of the ear can tolerate while still being able to identify the relationship of the newly adjusted ratios to the original simple ratios. No one who has heard a tempered interval and then heard its true equivalent could consider this procedure entirely successful. Simultaneous frequency relations that were in simple ratio are now precariously out of focus. The peak resonance of simple ratios, the very basis for conceiving harmony, is weakened.

Two popular forms in which temperament has been proposed for utilizing twelve steps to the octave, are mean tone temperament and equal temperament. Mean tone temperament came into use about the beginning of the eighteenth century. Equal temperament, although it was proposed even before mean tone, did not generally supersede it until the middle of the nineteenth century. Equal temperament is presently used. The twelve steps are based on the octave ratio, or $2/1$, by making each step equal to one another as each step is equal to the $\sqrt[12]{2}$. Only the

From C	Note	D	E♭	E	F	G	A♭	A	B	C'
	Interval	M2	m3	M3	P4	P5	m6	M6	M7	Oct.
Pythagorean (mode from C)		$\frac{9}{8}$		$\frac{81}{64}$	$\frac{4}{3}$	$\frac{3}{2}$		$\frac{27}{16}$	$\frac{243}{128}$	$\frac{2}{1}$
Natural (CM & Cm's)		$\frac{9}{8}$	$\frac{6}{5}$	$\frac{5}{4}$	$\frac{4}{3}$	$\frac{3}{2}$	$\frac{8}{5}$	$\frac{5}{3}$	$\frac{15}{8}$	$\frac{2}{1}$
Equal Temperament semitone $= \sqrt[12]{2} = \frac{105946}{100000}$		$\sqrt[12]{2^2} = \frac{112246}{100000}$	$\sqrt[12]{2^3} = \frac{11892}{10000}$	$\sqrt[12]{2^4} = \frac{125992}{100000}$	$\sqrt[12]{2^5} = \frac{133484}{100000}$	$\sqrt[12]{2^7} = \frac{14983}{10000}$	$\sqrt[12]{2^8} = \frac{15874}{10000}$	$\sqrt[12]{2^9} = \frac{168179}{100000}$	$\sqrt[12]{2^{11}} = \frac{188774}{100000}$	$\frac{2}{1}$

Fig. 1

Note From A3	B 3	C 4	C#4	D4	E4	F4	F#4	G#4	A 4
Interval	M2	m3	M3	P4	P5	m6	M6	M7	Oct.
Natural (M's & m's) in c.p.s.	247.5	264.	275.	293.3333	330.	352.	366.6666	412.5	440.
Freq. diff. that causes beats (sign is E.T. error in c.p.s.)	— .5584	— 2.3744	+ 2.1826	+ .3315	— .3724	— 2.7718	+ 3.3278	+ 2.8047	0
Equal Temperament (from A3=220 c.p.s.) in c.p.s.	246.9416	261.6256	277.1826	293.6648	329.6276	349.2282	369.9944	415.3047	440.

Fig. 2

2/1 ratio is exact. All other intervals are incorrect by varying degrees. Mean tone temperament allowed certain scales to be less in error than equivalent equal temperament scales, but the remaining mean-tone derived scales introduced greater errors. Some of the intervals in the worst cases beat so badly as to be called the "wolves". All equal tempered scales are at least in error equally. If the ear can tolerate one, it can tolerate them all. All eleven modulated positions from any note are now equally suited to act as the fundamental of a scale. Also, the scales may be modulated with equal degrees of resonance from the resulting frequency relations. These resulting ratios are of a high order. For example, the 3/2 is now about 14,983/10,000, or $^{12}\sqrt{2^7}$, as shown in Figure 1. That the intervals do not so neatly fall into a logarithmic division and that beats do arise in the audible range can be seen from a study of Figure 2. In Figure 2 an octave in the middle of the keyboard was chosen since it is the area most frequently used. After having been seduced by temperament a couple hundred years ago, it still appears that there is no better way to have scales and still modulate.

An obvious alternative is to increase the number of intervals within the octave and then to temper them in a similar fashion to that used now. One novel form proposed is to divide the octave into 53 equal parts or "commas" as originally suggested by Nicholas Mercator in the seventeenth century. But an error still exists when intervals taken from the 53 subdivisions are compared with the true interval ratios. Although as you increase the number of tempered intervals to the octave you reduce the error in relation to the original simple ratios, you will never reach the point where a tempered interval is equal to a true one. The number of intervals necessary to be seriously capable of fooling the ear is so large as to truly be called obese. Temperament is never exact; it is a compromise.

An interval can be constructed in two ways. It can always have one frequency as the fundamental or it can be the difference between the two steps, neither of which are the fundamental. For example in the C major scale, F is a perfect fourth, or 4/3, from C, the fundamental. F can also be a minor second above E, that is, 16/15, or a major second below G, or 9/8, etc. Comparison by steps illustrates the uneven frequency spacing between adjacent notes, which is one facet of a natural scale that equal temperament has smoothed over. For example, equal temperament recognizes only one minor second but there are actually several where natural scales are concerned: 16/15, 21/20, 25/24. Also, the minor seventh can be either 7/4 or 9/5; the tri-tone 25/18, 7/5; etc.

2
Theory of the New System

Have you ever asked yourself in considering the evolution of music: "what is music trying to do?" For example, scales of fixed pitch instruments were meant to advance and find a way to modulate. Music as a whole has had to evolve to meet the requirements of this major faction of the orchestra. Humanity has a way of covering over the wound. It would seem that in this fashion, the modulating scales that introduced unwanted additions of dissonance in turn anticipated the higher levels of dissonance that could now conceal the imperfect consonances of a tempered system. The modulated scales were first transfigured into the exploring art form using wholetone scales and finally probed towards keyless or twelve tone music. Artists of every generation seem to be attempting increasingly to remove themselves from the equal temperament problem. If you accept this trend, or even part of it, as natural, does it seem too late to replace equal temperament with anything short of total abolishment of frequency ratio relationships? Or are you an ultra-conservative who merely wishes that the guide lines of music would stay put in one key if that is the only practical way to make harmony sound best?

Man strives for a more sophisticated art. In music based

n fixed pitches this is an art that includes modulation, reely uses scales other than the diatonic, moves totally out of key centers. This is what music is trying to do. Is it possible to remove the temperament from under all of this and to come up with a superior substitute? A system that is reasonable and all inclusive while retaining simple ratio relations?

Let us start approaching our ideal by first clearing our minds of what we would normally consider a fixed pitch instrument to be, that is, one comprising a fixed set of tone generators. Let us continue by freeing ourselves of any avenues traditionally blocked to our inspection. One such avenue is the customary relation of all frequencies to a given fundamental standard. We shall freely pick the first note's frequency. This is now our chosen standard. Does it really matter specifically what that frequency is? Have not composers worked by hundreds of *standards* in the past? Today we perform music composed before our present 440 c.p.s.—A_4 standard. The transposing error can be considerable. Does it really matter? Even if this is a malpractice, it is generally unavoidable. So in picking a frequency, let us choose what we hear and record it with reasonable accuracy. That the first frequency should be absolute should not concern us as much as its relation in conjoining it with succeeding frequencies.

Music essentially moves in time either horizontally for melody or vertically for harmony. Therefore, the composite act of a composition can be broken down for observation to a single step no matter how complex the composition is or how much is taking place simultaneously. Let us move from our chosen first frequency by a single step. This step will be in ratio to the first note. It will be more important to make our ratios exact when simul-

taneous notes are occurring so that no greater levels
dissonance than the composer originally intended w.
occur. But just on the principle of keeping the system
wholesome, it can as well be an exact ratio for a melodic
step.

What has happened is that we have provided a frequency for our origin, considered it as an input, converted
it to our next frequency by a ratio, and received at the
output a new frequency that is now ready to be the next
input in our chain of musical events.

Again the melody is advanced by another step, or ratio,
as the previous note is changed from an output relation
into an input. It goes into the ratio converter, which is
set to any ratio we choose, and comes out the new note.
In constructing the foregoing three notes we have moved
from f_1, our selected fundamental or starting point, multiplied it by a ratio to get f_2, and multiplied f_2 by a ratio
to get f_3. Remembering that if f_1 were raised up a scaler
fifth in frequency it would be multiplied times 3/2; likewise, if f_1 has been lowered a fifth, it would be multiplied
by 2/3 since 2/3 is the inversion of 3/2.

Supposing at f_3 we wish to include other simultaneous
notes. To do this we will figuratively pause in time and
proceed to move outwardly in a vertical direction. At
this point we have chosen to place two frequencies over
f_3 to construct what can be considered as a scaler derived
triad, that is, f_3 times a ratio equals f_4 times a ratio equals
f_5. But since we have decided to continue our composition
from f_3, which is the origin note for both our chord and
our succeeding melodic step, we return to the frequency
f_3 and multiply it by a ratio to get the adjacent step to f_3
which will be f_6. We choose to continue melodically from
f_3 to f_6; however, we could also have continued melodically
to f_6 from either f_4 or f_5.

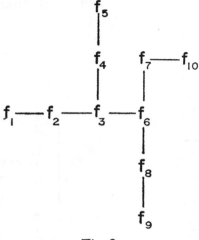

Fig. 3

For the purpose of demonstrating our potential flexibility, let us again extend vertically at f_6 so as to place f_7 above and f_8 to f_9 below f_6. To do this we move in our chosen ratio from note f_6 to f_7. Then returning again to frequency f_6, we move from f_6 to f_8 to f_9. Now we might choose to move off melodically from what we can compare to a scaler four note chord, to our next frequency, f_{10} via f_7, etc. The motion of our ten ratio-related notes can be shown with reference to Figure 3.

We have moved both melodically and harmonically in simple ratios. In so doing we have preserved the exact simple ratio relationships of connected steps. This is of fundamental importance to music that previously would have to be limited to a single scale of fixed frequencies in order to provide equivalent simple ratio harmony and melody. By this new means, modulating is not limited since there are no scale steps of fixed frequencies present any more. A group of fixed frequency relationships can be constructed as they would be if based on a natural scale and then be freely moved to be based on another scale without requiring the conventional pro-

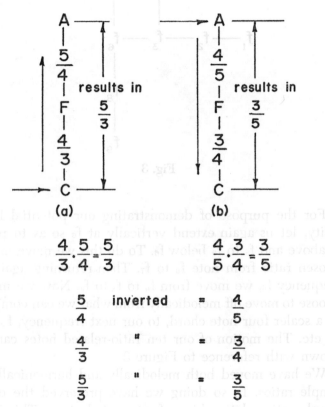

$$\frac{4}{3} \cdot \frac{5}{4} = \frac{5}{3} \qquad\qquad \frac{4}{5} \cdot \frac{3}{4} = \frac{3}{5}$$

$$\frac{5}{4} \text{ inverted } = \frac{4}{5}$$

$$\frac{4}{3} \quad " \quad = \frac{3}{4}$$

$$\frac{5}{3} \quad " \quad = \frac{3}{5}$$

Step-by-step—from bottom up or top down.
Ex: FM chord in 2nd. Inversion.

Fig. 4

vision of a new set of fixed frequencies. As will be shown later on, there is no limit in the building up of a harmony. Contemporary harmonies founded on the equal temperament system, such as polychords, major-minor chords, four- and secundal-chords, can operate as complex harmonies. Now they can also be constructed in simple ratio relations. In so doing, the unattended problem of large potential dissonances drowning out a complex harmony based on weak consonances of a high ratio order, which occurs in the equal temperament system, can be eliminated. Moving in steps can be performed with or without relation to a conventional key center or centers.

However, before getting too far ahead in understanding what will be going on, let us first analyze this new system from its basics and see what will constitute the new ground rules.

Harmony has been both the rose and the thorn in the last few centuries of musical development. Therefore, with due respect to established and new harmonic considerations, let us start by working forward from some simple ones.

If we take for our first example a chord such as an F major chord in second inversion, which is C-F-A, we find there is an option in how the steps of it can be assembled. In Figure 4, C-F-A is assembled from ratio to ratio; in Figure 4(a), from note C up 4/3 in frequency to note F and then up 5/4 in frequency to note A. The horizonal arrow shows which note is related to the last melodic step and therefore the starting point of the next chord to be assembled.

In Figure 4(b) C-F-A is again assembled step-by-step, that is, ratio by ratio. However, here the input is at the A, so that we move in the reverse direction from that in

Figure 4(a), downward. In both examples, the outlying notes C and A, the interval of a major sixth, result in a correct simple ratio relation of either 5/3 or its reciprocal 3/5.

In Figure 5, an alternate method is presented for assembling the same C-F-A triad. However, regarding Figure 5(a), F is found by its direct relation to C and likewise A is found by its direct relation to C. Now we are working

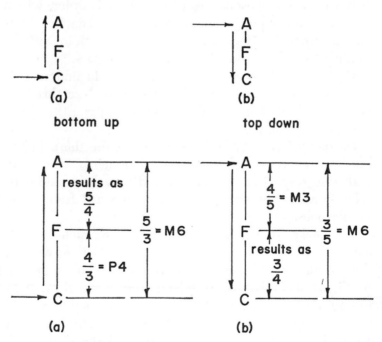

Always from the "fundamental" or end note. Bottom up or top down.
Ex: FM chord in 2nd Inversion.

Fig. 5

from the bottom note up, and the ratios for assembly are always taken from other notes of the chord to the bottom note rather, than as in Figure 4, between successive steps.

In Figure 5(a) C is the input and F and A are assembled above it. In Figure 5(b) A is the input and F and C are assembled below it. In Figure 5(a) we work from the bottom up, which is nearer conventional practice than going from the top down, as in Figure 5(b). In (a), F to A, or 5/4 is a result of the assembly of the two other ratios; in (b), C to F, or 3/4, now becomes the result of the alternate operation. In this case ratios are now *not* multiplied by succeeding ratios.

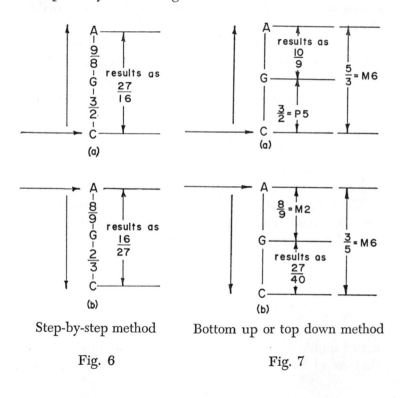

Step-by-step method   Bottom up or top down method

Fig. 6   Fig. 7

Comparison of the four means given for assembling the C-F-A triad in Figures 4 and 5 shows that identical results occur.

In most cases of triads both the step-by-step approach and the bottom-up (alternately, top-down method) will work. Figures 6 and 7 show an example where both will not work. Comparing the results of both figures should make this clear.

In Figure 6(a), going upward step-by-step results in the C to A being 27/16. C to A is a major sixth, or 5/3, in its lowest ratio form. The deviation is 1 part in 48. In Figure 6(b), going downward step-by-step results in the C to A being 16/27. C to A is a major sixth, or 3/5, making the deviation 1 part in 135.

In Figure 7(a), going from the bottom up, or C to G, which is 3/2, and then C to A, which is 5/3, results in the G to A becoming 10/9. G to A is a major second which in its lowest ratio form is 9/8, making the deviation 1 part in 72.

In Figure 7(b), going from the top down or A to G and then A to C makes the G to C result as 27/40. The G to C is a fifth which is 3/2, making the deviation 1 part in 120.

The example of C-G-A is not a common triad but it can occur easily enough. Therefore, it must be anticipated. It could be considered as the first inversion of the four note A minor seventh chord, that is, A-C-E-G with the E deleted.

We are approaching two significant matters that should be explained before proceeding further.

(a) Moving from step to step in music by ratios is the act of multiplying simple ratios by one another. We find that as in the Pythagorean system, multiplying simple

ratios in a sequence can produce an increasingly high
order ratio. Curiously enough, this matter has no real
effect on the melody. It would only have a bearing on
the melody were ratios to be freely extended from the
fundamental in a step-by-step manner so that the fundamental relationship to any given step had to be preserved
in simple ratio.

While at this point we wish to consider all aspects of
melodic movement, step-by-step movement should serve
predominately. In step-by-step motion we are comparing
one frequency to a succeeding one as a ratio. Since we
only hear one melodic note at a time and must apprehend
a sequence of related melodic notes as successive intervals, the process is a natural one. The alternative to step-
by-step melodic movement extending from the fundamental frequency is similar to the bottom-up method for
building harmony as shown in Figure 7. Given the first
five melodic steps of a melody, f_1, f_2, f_3, f_4, f_5, then f_1 is
related in simple ratio to f_2, f_3 is related in simple ratio to
f_1, f_4 is related in simple ratio to f_1, f_5 is related in simple
ratio to f_1, etc. This would serve to duplicate, in a conventional sense, one natural scale without modulation. It is
a considerably less flexible approach and is meant to be
shown only in passing. Again, while the high order ratio
potential of step-by-step progression need have no real
effect on the melody, it is of considerable importance to
harmony, where there is a sustained comparison of a
simultaneous sequence.

(b) There is a contradistinction between the simple
ratios that certain chords are built upon with notes in
conventional relation to the predominant note of the
given harmony, and the higher ratios that are then arrived
at between the steps.

Comparing statement (a) with statement (b) we conclude that a melody can, if we so choose, progress step-by-step without regard to a fundamental frequency. A harmony must be built from the bottom step up in order to retain a coherence of most simple ratio relationships between the constituent notes. To cover all cases of harmony the bottom-up method of constructing is the choice that is free of problems arising from exceptions.

In Figure 7 the bottom-up and top-down procedures show identical top and bottom, or extreme, frequencies for a given input. However, the middle frequency is dependent on which procedure is used, and the resulting high order ratio falls either between the middle and top notes in Figure 7(a) or middle and bottom notes in Figure 7(b). It would seem most logical to be consistent and work from the bottom up. This would be the closest parallel to conventional practice. Also, it puts the highest order ratio involved higher in the frequency range where there is a better chance that it will add less to the overall level of dissonance. This would make the order of establishing notes the same regardless of which note in the chord is linked to the preceding melodic step. Referring to the C-G-A chord as assembled in Figure 7, if the input was A, the route is A to C, C to G. If the input is G, the route is G to C, C to A. If the input is C, the route is C to G, C to A. Being the bottom note, C in each case acts as the pivotal base point from which the other notes in simple ratio extend.

With reference to Figure 8 in the following harmony there appears to be no question but that the bottom-up route is preferred to the step-by-step. Step-by-step would make A to A' 162/80 which is a high order ratio that is close enough to 2/1 to provide a potential source of beat dissonance.

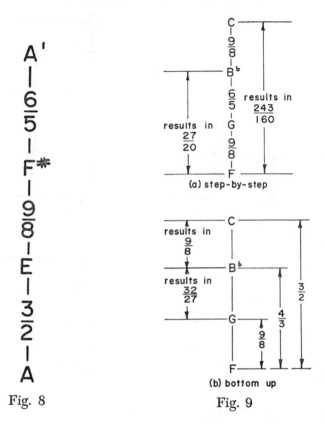

Fig. 8

Fig. 9

A similar problem arises in the secundal chord given in Figure 9. It should be noted that a 32/27 has resulted in the bottom-up procedure where G–B flat would ideally be 6/5. But the discrepancy would also be present in a conventional harmonic construction based on a scale of fixed frequencies. Also, in this example the step-by-step results in two high order ratios, F to B flat as 27/20, which ideally would be 4/3, and the F to C as 243/160, which ideally would be 3/2.

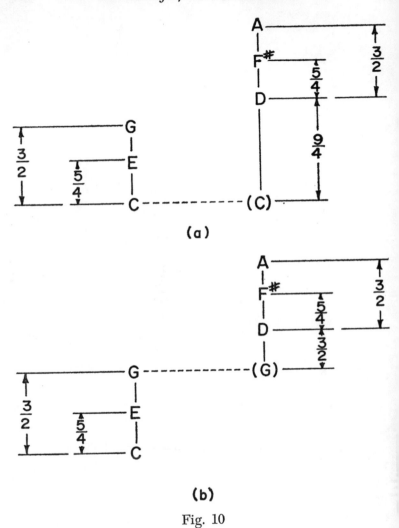

(a)

(b)

Fig. 10

As harmony has advanced it has also become increasingly complex. One approach to contemporary harmony is the polychord that involves paralleling of chords from two different keys at once. The polychord, a combination of two triads, can be assembled in two ways. Regarding Figure 10(a), the tonic of each triad links the triads, which appears to be most natural. The alternative in Figure 10(b) links the fifth of the C major triad to the tonic of the D major triad to relate both as a continual ascending progression. Although the two relations within the polychord are both possible considerations, what is actually occurring is a single combined harmony. A chord of six notes is occurring, which the ear may be able to

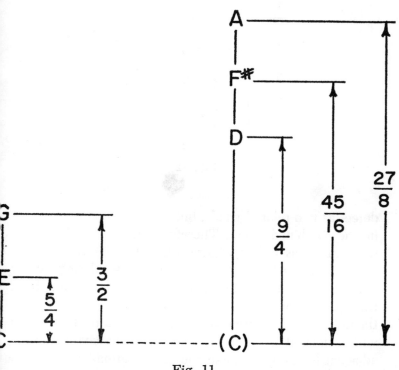

Fig. 11

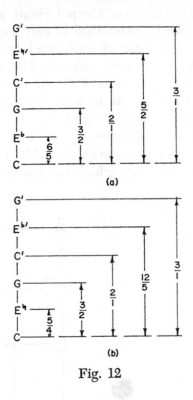

Fig. 12

detect as two related triads, but which interact and result in one combined effect. Therefore, Figure 11 shows the same ratios as either Figure 10(a) or (b) but as they would *actually* interact. We are now considering all the note relations as one continuous sequence related from the bottom note up. It would seem that in this example the polychord presented is inevitably like a thirteen chord with a deleted seventh, or, vice versa. It remains for the adjacent harmony to make the classification.

Even accepting some of the higher ratios in Figure 11

than were used in the original triads, the polychord was never this near to maximum consonance ratios when based on the higher equal temperament ratios.

Our next example is the major-minor chord. In Figure 12 two triads taken from the equivalent scaler key of C are used, both in root position. The construction is from the bottom up. In Figure 12(a) the minor chord is below the major and in Figure 12(b) it is above. In Figure 12(a) neither the minor nor the major triads present any problems. Noteworthy is that the E′ natural from the major triad is related to C as a 5/2, which is 2/1, or an octave up, times a 5/4, which is exactly what a major third should be in lowest ratio. This means that the E′ natural results in both maximum consonance to the C′, or root of its triad, as well as relating in maximum consonance to the C, which is both the fundamental of the six note construction and the lower triad. In Figure 12(b) equally suitable relations result.

So far several configurations of harmony have been considered but in each case with only a single melody line connecting them. Several melody lines occurring simultaneously, or counterpoint, is essentially a process for working within fixed scales, that is, where each note has a fixed pitch. The melody lines have primary importance and the harmonic relations that result are secondary. This is not to suggest that the resulting harmonic relations are insignificant, for if not controlled, they can combine in a progressing passage in quite a haphazard manner. A composition is normally bounded by levels of consonance and dissonance. This is a manner of delegating certain harmonic relations as being acceptable in the piece while deleting others as being too far removed or capable of altering the harmonic level of the composition. These categories are what can distinguish a composition as belonging to a certain period in music. Therefore, returning to the relative harmonic level that can result as a secondary consideration in counterpoint, it becomes increasingly evident that any worthwhile composition in counterpoint does not have its harmonic relations neglected. The harmonic relations are quite important as they are in actuality the *plurality* of melodic movement at a given instant.

No matter how complex the contrapuntal relations can become, what is heard at a given instant is the compounded sound, or harmony, of all the constituent notes. Counterpoint can even be considered as essentially a group of harmonic progressions. It can be distinguished from a single melody line with included harmony by the fact that each voice proceeds with a greater continuity with the melodic line and that the quantity of notes present at a given time remains generally more constant. This consistency, and frequently the different instrumental timbres involved, allows the observer to be able to separate and follow a single voice.

Since the primary purpose behind a musical system maintaining simple ratios between notes is for maximum consonance and clarity of simultaneous sounds, attention to harmony should be upheld as a paramount factor. This would appear especially so in the light of contemporary approaches to counterpoint where more than one key may be in operation at once and where the voices often have harmony attached to them. Where harmonies are extending from melody lines in different keys, complex chords can result that require attention in order to be reasonably related in simple ratios.

A short musical illustration in counterpoint is provided

Fig. 13

in Figure 13. In bar 1, both melodies are in the scale of G major. In the second bar, Theme B modulates to the key of G♯ major. However, the key of G major continues for Theme A into the second bar. Therefore in the second bar, G major and G♯ major are in complementary relation. Assembling this musical segment as a harmonic relationship presents no problem and allows the two different keys to meet in the second bar while still retaining simple ratio relations between the ensuing harmonies.

As was previously found in the case of harmony, including multi-keyed chordal relations, the bottom-up method is most suitable. Therefore, in trying to provide an approach that benefits the maximum overall potential usage, the bottom-up assembly is best used throughout counterpoint.

The assembly of the counterpoint of Figure 13 is shown in Figure 14 as constructed by harmonic relations. In this example Theme B consists of the bottom notes. The assembly of Theme B moves from step to step and at each step, working from the bottom note up, a harmonic

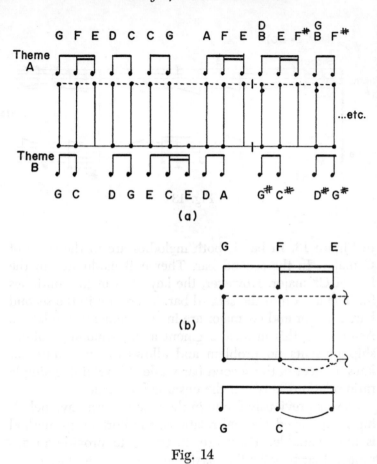

Fig. 14

relationship is constructed from Theme B up and through Theme A. If Theme B had harmony in conjunction with its melody line, then regardless of what note of a particular harmony was the input, the given harmonic relation would start from the bottom note up; this is in accord with previous examples of the assembly of harmony.

Therefore, with reference to Figure 14(a), the first note, G, of Theme B is related in simple ratio to the first note in Theme A, G, two octaves up. The note G of Theme B is then related to the next note in the melody which is C, in step-by-step assembly. C is a fourth above G, or 4/3 of the first or fundamental note. C of Theme B is now

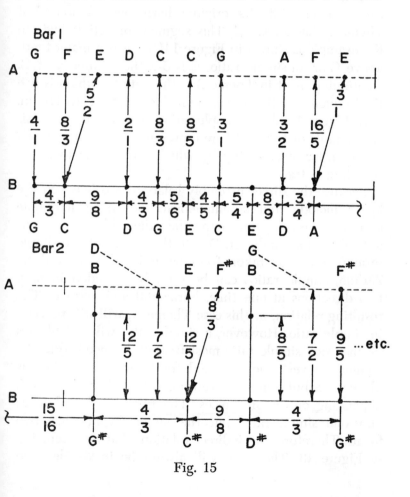

Fig. 15

related in simple harmonic relation to the note F in
Theme A. F is an eleventh above C, which is an octave
plus a fourth, or 8/3. The next note to appear in the musical progression is the note E in Theme A. Working again
in a consistent manner from the bottom up, E of Theme A
is related to C of Theme B since the note C is still being
carried over while its original harmonic relation F of
Theme A has subsided. This segment of activity within
the example is shown in Figure 14(b). E is a major tenth
above C or, in simple ratio, E is 5/2 the frequency of C.
Continuing, next in Theme B is the note D, which is 9/8
the frequency of the preceding melodic note . . . and so on.

In Figure 15 all the simple ratios involved in the musical example of Figure 13 are represented. They are shown
in the order of assembly by which they relate to corre-
sponding notes.

Because the direct assembly relations are between the
lower melody and simultaneously occurring harmonic
relations, the remaining upper melodic relations occur as
a result and are indirect. Due to the nature of the simple
ratio relations as taken from natural scales derived by
Zarlino, simple ratios can be directly extended in only
two directions at one time, horizontal and vertical. The
resulting relation, in this case Theme A, will still proceed
in simple ratio. However, a simple ratio will not always
be the exact simple ratio most frequently specified as the
common lowest one accepted for an interval: a fifth is
always 3/2 but a minor seventh can be either 7/4 or 9/5,
depending on circumstances. The fifth is referred to as
only one simple ratio but the minor seventh has two
forms. Therefore, the indirect relations that are occurring
in Figure 13, Theme A, will always be in simple ratio

from step to step, but not always the one in common usage.

Again referring to Figure 15 and starting from the beginning of the musical phrase, G of Theme B to the C of Theme B to the F of Theme A is 4/3 times 8/3 which makes F 32/9 in frequency above the G. Taking an alternate route, we can go from the G of Theme B to the G above it in Theme A that is in direct relation and then go from the G in Theme A to the F following it in Theme A by indirect relation. We can take the G to F relation to be 8/9, although as an indirect relation we are now checking to see if 8/9 is present. Considering the two steps of the preceding operation together, one is a direct relation, one an indirect. This results as 4/1 times 8/9 which is 32/9. In this case the indirect relation, or the 8/9, has proven to be an equivalent to the exact frequency relation from common usage as would be chosen for the step-by-step motion from G to F, which is a major second interval in the first two notes of Theme A. The two assembly routes chosen, one valid because it is direct, one not so valid because a part of it is indirect, result in the same overall ratio, or 32/9.

Proceeding further in the musical phrase, C of Theme B to E of Theme A is 5/2 in direct relation. C of Theme B to F of Theme A is a direct relation but the next step, F of Theme A to E of Theme A is indirect. C to E is 5/2. C to F to E is 8/3 times 15/16 which results as 5/2. Again, the indirect relation is that of most common usage, or 15/16.

Following are direct relations C to D in Theme B up to D in Theme A, or 9/8 times 2/1, which is 9/4. Alternately are the direct then indirect relations of C of

Theme B to E of Theme A to D of Theme A, or 5/2 times 8/9, which is 20/9. Since 20/9 is not equal to 9/4, a deviation is present which can be expressed as 1 part in 36. However, the indirect ratio for E to D of Theme A is 9/10 which remains a simple ratio and one that is very close to the ratio of 8/9, which is commonly provided to describe the interval of a major second. Also it should be mentioned that the foregoing has resulted as the intended preference of keeping the harmony in optimum simple ratio since even a slight deviation to higher ratio in a harmonic relation can be much more easily detected due to the notes occurring simultaneously.

Up to now where there has been occasion, a certain practice concerning large intervals within a harmonic relation has been used and should be commented upon. When an interval between two notes must be expressed in simple ratio as the result of multiplying two smaller ratios, the constituent intervals should each be in lowest possible ratio. Such is the case in expressing intervals greater than the octave where relation to scales or intervals are normally expressed within an octave. The interval over an octave should be divided by the number of octaves that will go into it. The resulting ratio is multiplied by the remaining frequency difference expressed as an interval in lowest ratio. If you use the octave as one of the ratios, the lowest possible ratio will *always* result from the multiplication.

3
General Applications

Endless comparisons can be made between this new system of music progression and conventional technique. The results as have been briefly developed in preceding comparisons should be received as quite a bona fide improvement. Then again, as with anything novel, some details may yet have to be worked out. Perhaps more important than the enjoyment of bringing potential improvement to accepted, current technique, are partaking in the seemingly endless musical evolution and finding application in the most current thought and in the frontier thinking into tomorrow.

Although currently twelve tone music is hardly considered a frontier, if applied to the new step motion it must be freshly considered. It would appear dangerous to stereotype twelve tone music or even consider it historically as an exhausted medium until more time has elapsed in which conclusions may be more assuredly drawn. However, at this stage certain guiding principles are evident by their frequent occurrence. Essentially categorized, twelve tone form resists assimilation into any prior type of scale or harmony but rather manipulates a *tone row*, or order of notes, selected by the composer from the twelve tempered tones available within the octave.

Normally to further safeguard against similarity with any prior art, no particular tone within an order is repeated until all the other eleven tones have been used. Harmony in the conventional sense, is usually excluded, and when notes occur simultaneously they carefully avoid being compounded into anything that might resemble a harmony constructed from a conventional scale outline. Variety is achieved mainly through varying the tone row by inversion of the notes within the set or reversing the row in *crab-style*.

The twelve tone system can be considered as a logical outgrowth of equal temperament because equal temperament favors no key and abolishes simple ratios that are the foundation of scales. Equal temperament is not really even related to prior musical practice except that the intervals created are sufficiently proximate to be assimilated with the intervals of simple ratio established in prior practice. Because tempered intervals are actually *artificial* simple ratios, it serves merely to emphasize why a *puritanical* style would eventually evolve from temperament that was not an artificial carry-over, a style that was void of adjusted scales and scaler harmony relations. Twelve tone music applauds the even spacing between intervals of equal temperament and directly relates itself to equal temperament. Of any prior music form, it is actually the most clearly related to its tone system. How can a system intimately related to equal temperament be applied to simple ratios? In a direct sense, it cannot.

Previously, simple ratios took no other form than assembly into sacrosanct scales or modes. It was a practical necessity. In this new system of step motion from interval to interval, as previously illustrated, relations that are derived from scaler relationships are only one manner of application. Considering step motion of intervals in the freest sense, scaler relations need have no bearing. The system may be considered as independent of that prior art.

Therefore, let us start a new assessment of twelve tone music as simply twelve tonal divisions to the octave. The divisions are no longer equally spaced like equal temperament logarithmic relations to the base 2. The *tone-row* no longer hovers in a disciplined manner over geometric grid lines of duodecimal octave subdivisions. The *tone-row* now is an ordering in step to step motion by intervals and the intervals truly have no relation to a scale. There is no concern for a fundamental frequency as if it need be a universal standard. The first note is a personal choice of the composer and might never again appear as the exact same frequency. Harmonic possibilities are present but are now unlimited. There is no need to make a relation to any prior practice if this is the composer's intent. A harmonic relation need have no bearing on a previous harmonic relation. The only repetition is that intended by the composer as he controls the order of the interval relationships. Now dissonance can be abolished where not intended since simple ratios are still intact in the system. Simple ratios can be made to work in a non-scaler medium. Twelve tone music can obtain a level of concord previously lost as a part of its operation.

By further evolution twelve tone music no longer has to be a system of composition based on the compromise system of equal temperament that resorts to a manner of interval distortion that originally served as a crutch to gain new freedom to modulate and to explore new scaler possibilities. Twelve tone music exists now in its own right and can be looked upon as a direct descendant of Pythagorean philosophy, rather than as an offshoot of a historical compromise.

Why stop at a system that will indirectly have twelve subdivisions to the octave, that is, an interval within it of never less than a scaler equivalent of the minor second? Why even bother to fix or keep track of the number of different intervals that are present? There is no reason to limit the composer's choice, and the only practical limit is the audible separation of steps in frequency. If a small step is perceivable and desired, it can and should be used. Such thinking has been with us for a while. The present availability of compositional means can make such a dream practical.

And the next step is to ask why we should bother to follow the past in working in an equivalent manner that results in dividing a given octave of the audio spectrum into the same number of subdivisions as all other octaves. The human ability to distinguish small intervals, or frequency change, varies in the audio spectrum. This ability can be related to two factors. If we divide the audio spectrum into octaves, by way of illustration, any given interval in one octave will double in frequency an octave above, halve in frequency an octave below. Secondly, the ear is an asymmetrical device, or organ, that has a tuned bandpass to amplitude variations of air pressure. The bandpass is peaked around 3 kilocycles and slopes off asymmetrically to either side, regarding a just noticeable difference in frequency or amplitude. The asymmetrical slope is steeper towards the base response direction.

From the above it seems reasonable to deduce that there is no particular advantage to the ear in having a system based on, or related to, a fixed number of sub-divisions per octave. It can also be considered as wasteful.

In a general sense two approaches may be taken. When considering the application of an interval from the audio spectrum, the location in the audio band is first noted. The location is then generally related to one of the following courses.

The first is that the smallest interval that may be used will decrease in size as the frequencies of the audio band rise. This can be expressed another way: the number of different related intervals within a given octave, where one frequency of the interval is one of the frequencies of the octave, will increase in relation to the rising frequency area used in the audio spectrum. That is, there are more intervals per given octave as the frequency rises.

The second general approach, and perhaps the more relevant of the two, is that the smallest interval used will occur at the peak, or 3 kilocycle point, of the audio spectrum and to either side will slope off into gradually larger intervals governing the low order limit of intervals used. Another way of expressing the relation is that the number of related intervals used within a given octave, where one frequency of the interval is one of the frequencies of the octave, will be at maximum for the octave straddling the audibility peak. There will be a gradual asymmetrical slope off in quantity of intervals per given octave to either side of this peak, that is in the direction of rising or lowering frequency.

Again, the foregoing is meant to be considered as only a very general guiding rule for the composer. Its interpretation should prove to be a natural one since the limits of interval variety per location on the audio band are natural. If the composer cannot hear a change at any given point in the frequency band between two frequencies, then the practical limit is reached and has been naturally imposed. If the composer bases his technique on dividing the octave into intervals, he is now free in his practice to shift at any point in the music to a greater or lesser number of new intervals. Alternately, throughout the work he can consider himself as being completely free in his choice of intervals, not having to work from any particular set.

It is apparent that the choice of intervals can be restricted to *any* set number of divisions per octave. This will make previously proposed systems such as 19 tone, 24 tone, 31, 53, etc., equally accessible. At the same time it will free these systems from any prior dictates of being bound to a temperament. Increasing the number of intervals over the twelve per octave, thereby decreasing the minimum interval present, introduces successively higher order simple ratios. This would seem only to emphasize the need to eliminate temperament when dividing the octave into a large number of fixed intervals if any clear identity is to be retained for these intervals at all. A very small interval will reach the point where the frequency change is so slight in relation to adjacent intervals as to be near, or equal to, the errors introduced by a temperament.

4
New Pythagorean Application

Moving into the purely experimental area of application gives us the opportunity to re-examine the past and possibly re-evaluate it in the light of present understanding.

There seems to be little question as to the validity of presenting intervals in lowest and simplest ratio in order to provide the maximum consonance so that, regardless of the relative dissonance that is intended, no more will result than is actually desired. In the Pythagorean scale, which is the origin of modern music, we know that each note is derived from the exponential relationship of combinations of the octave, or 2^n, and the fifth, or

$$\left(\frac{3}{2}\right)^n$$

The result can be expressed as a ratio resulting from the required ascending fifths and descending octaves or vice versa. As the exponent increases, the resulting ratio becomes increasingly complex. This led Zarlino to reduce certain ratios, such as the major third, from 81/64 to 5/4 in order to put the *problem* intervals in lowest ratio. This we already know.

However, while subscribing to the favoring of lowest

53

simple ratios, Alexander Wood in his book *The Physics of Music* also presents a potential contradiction. He refers to an experiment conducted by Paul Greene (*Journal of the Acoustical Society of America,* Vol. 9, 1937, p.43), in order to measure the exact intervals produced by six violinists:

[There were] (a) marked deviations from the tempered intervals, (b) considerable variation in the extent of the deviations, (c) considerable agreement in the direction of the deviation, (d) a mean value for each interval which was nearer to that of the Pythagorean scale than to the natural scale. The last result is the most surprising. Thus in the case of the minor third we have [*author's note:* expressed in savarts; 300 savarts equal one octave]:

Pythagorean interval	73.5
Tempered interval	75.0
Natural interval	78.9
Average interval played	74.0

and in the case of the major third:

Natural interval	96.6
Tempered interval	100.0
Pythagorean interval	102.0
Average interval played	101.5

If this test could be validly applied to music in general, what might it prove? What would the results be if harmony were brought into this consideration?

We started by suggesting that an experiment might be in order. Let us return to that consideration to see if, just before the historic pivotal point Zarlino introduced, a new tangent of development can now be presented.

Equivalent interval of a scale, from C = fundamental	In exponential form	As a ratio
C ——— G	———	$\dfrac{3}{2}$
C ——— D	$\dfrac{\left(\frac{3}{2}\right)^2}{2}$	$\dfrac{9}{8}$
C ——— A	$\dfrac{\left(\frac{3}{2}\right)^3}{2}$	$\dfrac{27}{16}$
C ——— E	$\dfrac{\left(\frac{3}{2}\right)^4}{2^2}$	$\dfrac{81}{64}$
C ——— B	$\dfrac{\left(\frac{3}{2}\right)^5}{2^2}$	$\dfrac{243}{128}$
C ——— F#	$\dfrac{\left(\frac{3}{2}\right)^6}{2^3}$	$\dfrac{729}{512}$

Fig. 16

Again we are examining the Pythagorean scale. By way of illustration, let us ascend from our originating, or fundamental, frequency by six perfect fifths and then descend by the necessary number of octaves. We will establish a set of notes, or call it an experimental scale, within a given octave. This scale we can in turn duplicate in other required octaves. Our six note set is shown in Figure 16.

In Figure 16 each interval in our *scale* is related to the fundamental and is shown as it would derive exponentially and as the resulting ratio. The results are familiar. But this time let us concede the problem of increasingly high order ratios that are resulting, on the probability that something important may have been suggested by Greene's experiment.

If we choose to assemble a relatively extended scale of subdivisions for our preparatory octave, some frightening ratios may discourage any attempts at physical reproduction. In exponential form no high order numbers are ever present for there is only

$$\left(\frac{3}{2}\right)^n$$

and 2^n. The simplicity of the relation is possibly the key reason for advocating the development of the Pythagorean system as it stands.

Before digressing further into this area, it might be timely to examine the general approach to physical application. Up till now our ratio machine has been made to convert any input frequency into a new output frequency in accord with our selected ratio. Therefore, to produce an interval is essentially a one step process. Likewise, a triad is a two step assembly. Now let us prepare to extend the usefulness of our ratio machine by making it possible to reapply the output to another input and repeating as often as necessary. This time the frequency change by ratio will be computed for only one interval. Moving in several associated steps in order to get only one interval as the output will allow us to build exponentially any Pythagorean interval we so choose. This will remain so no matter how far we extend into the realm of high order ratios. It can now be accomplished with only two repeated frequency conversion ratios, the 3/2, 2/1 and their respective inverse relations the 1/2 and 2/3.

For illustration, let us move from the fundamental frequency, here taken to be the scaler equivalent of note C, to the note B. C to B can be expressed as the Pythagorean ratio of 243/128. It is also, exponentially,

$$\frac{\left(\dfrac{3}{2}\right)^{5}}{2^{2}}$$

It can be assembled with the two simple ratios 3/2 and 2/1 as: 3/2 times 3/2 times 3/2 times 3/2 times 3/2 times 1/2 times 1/2. Or it can be computed as the input frequency times 243/128. 243/128 is only one step in comparison to the alternate multiplication sequence requiring seven. However, each new interval will in turn be a new frequency ratio if considered as a single step. When applying the four ratios of the fifth and octave in the required sequence, there are no further high order ratio considerations. Therefore, our design favors the multiplication sequence since, as a whole, it is the simplest to physically create.

The ordering of the multiplication of the ratios has no effect on the output. For example, using the scaler relations, the major sixth of C to A can be either four fifths up and one octave down, or now alternately, three fifths up, one octave down and then a fifth up.

To be a little more analytical, with respect to the first example of deriving a Pythagorean major seventh interval above the fundamental frequency, the ratio conversion can be illustrated by the outline for this procedure in Figure 17.

The seven stages of ratio converters shown in Figure 17 take in twelve operations of individual simple fre-

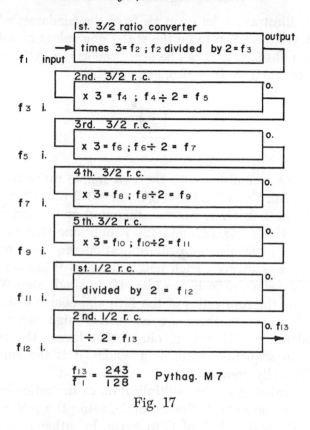

$$\frac{f_{13}}{f_1} = \frac{243}{128} = \text{Pythag. M 7}$$

Fig. 17

quency changes starting from the fundamental frequency input, which is f_1, and concluding as the final frequency output, or f_{13}. This results in the final frequency output being the Pythagorean major seventh interval above the fundamental f_1 input frequency. The first five ratio converters are identical. Also, they may be observed as dual functioning, that is, a multiplication and then a division takes place.

It is notable that the first five ratio converters are identical, as are the last two. Therefore, the output f_3 of the first ratio converter would most likely be, in actual practice, reapplied to the input of this same ratio converter four more times instead of using five redundant block units as shown in Figure 17. Between the output and the input of each repeat operation there would most likely be a memory present in order to hold the output long enough to clear the block ratio converter of the original input before reapplying the output. For example, f_3 would be held over in the memory, while the f_1 input was removed and then would be reapplied as the next successive input.

However, let us not dwell too extensively on an actual mode of operation at this point and for the time being, return to purely theoretical considerations.

What Zarlino provided in music evolution by reducing the high order Pythagorean ratios for certain intervals was the *wholesale* introduction of harmony.

For lack of better testimonial, let us ease our grip on the prime importance of simple ratios, especially when assembling a given harmony, and out of curiosity take sides with Greene in the controversial results of his experiment. We must also notice that in so doing, by leaving the ratio relationships, or intervals, in exponential form, the exponents of various ratios relate quite simply along side one another.

Take for example an A major triad in first inversion. Expressed in Pythagorean intervals from the bottom-up, we have, C to E or a major third which is 81/64 and C to A, which is a major sixth, or 27/16. Regarding the major third as a simple ratio, we would tend to see it as a higher order than most desirable. The major third would be rounded, or reduced, from 81/64 to 5/4. Likewise, the

major sixth, or 27/16, would be reduced to the low order
ratio of 5/3. However, this given chord expressed in
Pythagorean exponential form is C to E, or

$$\frac{\left(\frac{3}{2}\right)^4}{2^2}$$

and the C to A as

$$\frac{\left(\frac{3}{2}\right)^3}{2}$$

Placed in comparison, we have still retained only two
simple ratios. However, they are now exponential ratios
whose only difference is the power of their exponents.

As shown in the foregoing example of C-E-A as a
Pythagorean related harmony, the relationships are quite
simple if left in exponential form. In comparison to ratios,
the exponential form may prove easier to handle. Will it
be better sounding?

In applying the Pythagorean intervals to the new direct frequency-to-frequency relation system, all musical
possibilities previously considered for application to simple ratio intervals can apply equally well for the purpose
of comparison and evaluation. Our test palette consists of
modulating harmony, wholetone scales, multi-keyed harmonies, twelve tone and microtonal systems, and frequency response controlled microtonal systems. Everything that we have mentioned prior can be tried.
Comparisons that would previously have been impossible to make are possible with either the natural or tem-

pered systems. How far and how well will the Pythagorean exponential intervals compare with their rounded lowest ratio equivalents? Is the natural scale derived by Zarlino *most* natural? A working model will answer this controversy.

5

A Preliminary Physical Design

There are several ways in which a frequency can be shifted. In so doing certain basic requirements must be recognized. Of fundamental importance is that the frequency change be in the exact ratio intended. Otherwise, errors can possibly add up to deviations sufficient to nullify justification of the procedure. Accuracy is not difficult to incorporate as long as it is given paramount importance from the beginning.

A second fundamental consideration in the frequency shifting process is that the input frequency be variable and may be anything within the audio spectrum that a composer would use for a musical note. Since the input is variable, the output frequency will be also. This would immediately eliminate any manner of tuned circuit in the design of the frequency changer.

Another basic factor is that it would appear preferable to separate the frequency shifting process from the final wave shape section. The final wave shape would be the completed musical sound. The frequency converter will handle either a sine wave or a pulse, whichever best meets the specific needs of the design chosen. The frequency converter will then forward the pulse, or sine wave, to the actual music-generating source.

At the music-generating source several basic approaches are available. The pulse, or sine wave, can trigger a device that in turn produces a complex wave shape, one rich in harmonics. This is then shaped to produce the intended musical sound. The shaping would be a formant circuit. The complex wave generating device could produce waves with shapes that might, for example, be square, sawtooth, or triangular. If the complex wave shape desired was a square wave, the device used could be a Schmidt trigger. Wave shaping procedures are familiar to anyone involved in electronics.

The pulse or sine wave output of the frequency changer can also be converted into the required musical timbre by synthesis. The pulse, or sine wave, would trigger the necessary group of basic wave shapes making up the required harmonic series. They would then be compounded in their desired amplitude balance with one another to produce a complex wave envelope, so creating the desired musical sound.

Returning to our consideration of the frequency conversion, we again note that the input and output of our device will be continuously variable from operation to operation, that is, from musical interval to interval. A basic example of this procedure is in order.

A simple way to change a frequency in any ratio required is by driving an alternator with a synchronous motor. This would be an electromechanical frequency conversion. There are three basic elements in the timing: (1) the relation of the synchronous driving motor to the given input frequency; (2) the relation of the synchronous driving motor to the driven alternator; (3) the relation of the alternator to the output frequency. The resulting frequency change can be in whatever exact ratio is required. If a 60 c.p.s. sine wave were operating a four-pole synchronous motor, which is in turn driving a twelve-pole alternator, the resulting output will be a 180 c.p.s. sine wave. A 3/1 frequency conversion has taken place. If the input frequency were considered as that of a musical note, then the output would be one of a twelfth above it. Likewise if the alternator and motor are reversed in their function, the frequency will be in a 1/3 step down ratio.

The foregoing electromechanical frequency changer is quite similar to what takes place in the Hammond organ. In the Hammond the propulsion is by a synchronous motor driven from a 60 c.p.s. power line. In this application the motor is chosen for its ability to retain a constant speed for an indefinite length of time. In turn the motor is connected to a plurality of small alternators, each one related to the driving motor and to each other. This results in the production of twelve equal tempered semitones extending over the full organ manual. These sine waves are used in various combinations and their individual amplitudes are also controlled to synthesize both the fundamental frequency of a musical note and the tempered harmonics. While such a system has no direct relation to our interest, general design parallels are obviously drawn.

One potential drawback of electromechanical frequency conversion with a variable frequency input, is the time required between a given frequency change. The intended use of an electromechanical arrangement would primarily be for programming a composition at a con-

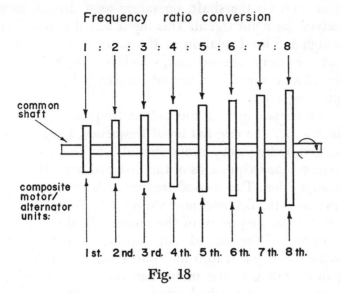

Fig. 18

siderably slower rate than playback. Therefore, the speed is not necessarily a primary consideration.

It would seem desirable, for simplicity, to mount more than a pair of driving and driven units on a common shaft. Also possibly the design could composite the individual units into the dual function of driving and being driven. Since simple frequency ratio conversion has always been the musical objective, the number of units required to obtain the desired frequency ratio changes would be at a minimum. By way of example, and with reference to Figure 18, driving and driven combination units can be attached to a common shaft, each in a consecutive ratio relation to each other in whole numbers from one through eight. If a given input frequency were changed in the ratio of 3/2, then the third combination driving and driven unit on the shaft, operating as a driving motor, receives the input signal. This input signal is transferred through the fixed relation of the units to their common shaft over to the second unit, which in turn performs the alternator function. The output frequency is now 3/2 the input frequency.

Now supposing that the output frequency just given, which is 3/2 the original input frequency, is again to be changed by frequency conversion of 3/2. That is, we have produced the frequencies of the first two musical notes of a composition. The second frequency is an interval of a fifth above the first and now we want to assemble a third note 3/2 the frequency of the second. To do so we take the output of the second unit, consider it as an input in function, and put it into the third unit. The third unit is again operating as the motor and takes the second note frequency in our musical sequence from the second unit, which is acting as the alternator. The unavoidable time lapse of our electromechanical frequency converter would be that required to accelerate the shaft speed since the output frequency of the first operation, now to be used as the second input frequency is 1½ times higher than the input frequency of the first operation.

The range of intervals of our hypothetical frequency converter can be either a step-up or step-down frequency ratio whose whole numbers are between one and eight, for example 4/3 or the musical interval of a fourth up, 8/5, which is the minor sixth up, 1/2, which is the octave down. However, an additional component is needed to make the ratio converter operative for more than one step. The required addition is a memory.

In the previous double operation example, the output frequency of the first operation became the input frequency for the second operation. This would mean that inputs for both the first and second operation would be simultaneously presented to the third frequency converter motor unit on the common shaft. This would not work since there obviously must be a delay between operations, or a cut off time, where the first operation is removed and the second one applied. However, the output of the first operation must be retained for two reasons. It must be held as the second note's frequency for later playback of the musical sequence at the intended rate of final performance. Also, it must be able to act as the input signal for the second operation even after the input generating source, or first frequency, has been removed to clear the unit for the next operation.

For these two reasons the output of the first operation is retained in a memory. Referring to computer design where memories are an essential part of their overall function, a memory drum would appear a most likely choice for the requirements. A second choice for the memory that could be applied to our requirements would be magnetic discs. There are other types of memory storage units but, differing from the normal requirements of a computer, our message is a continuous frequency rather than bits, units of a binary code. Our memory would also have to be non-destructive, that is, a memory that can write/playback the prior read/recorded message without losing it. Magnetic memory cores are destructive and are an example of what would most likely not be applicable. Also, magnetic memory cores are not meant to handle a continuous frequency. Time is regarded in our machine as a constant and our memory device must be one having an accurate means of regulation for repetition and exact sequential frequency relation.

Where a memory drum, or disc, is used, the rate of rotation would be considerably slower than is common practice for computer use. In our application, which would generally favor using as large a diameter drum as practical, one revolution would take at least several seconds. In computer usage, rotation speeds from 1200 r.p.m. on up are common.

Briefly, one basic type of memory drum construction is a non-magnetic cylinder coated with a layer of iron oxide over which a layer of lacquer is applied and then buffed. This cylinder is motor driven and revolves past record/playback heads situated around its outer periphery. It is normal to have one record/playback head combination per track with more than thirty tracks per inch spaced along the length of the drum. The drum diameter for computer application usually ranges from four to fifteen inches. For our requirements fifteen inches would most likely be the minimum diameter feasible unless the peripheral movement of the drum past the heads is considered to be less than a speed of 7½ inches per second. Seven and a half i.p.s. is arbitrarily chosen as equivalent to that used in most conventional tape recordings of music. However, if the wave shape of the ratio converter is to be only a sine wave or pulse, the spacing of the magnetic fields on the tape could be closer than needed for actual high fidelity. Therefore, a much lower speed might be feasible in the area of about 1 7/8 i.p.s. In turn the required diameter of our memory drum is lowered.

Since there can be thirty or more tracks per inch on a computer memory drum and since there is a record/playback head for each track, the heads are staggered around the periphery of the drum in order to provide clearance for the heads.

Each track of the memory drum is perpendicular to the axis of the drum's rotation and may be likened in operation to tape loops. However, where each track in a computer application holds numerous *bits* of information, each track for our musical application will be for a single frequency message. For our purposes two record/playback heads might be sufficient, each of which can be positioned by suitable mechanism over the required tracks in use. Alternately and most likely, more than two heads will be used at a time in playback for the final recording. The series of heads required would be gang-mounted, like computer practice, so that each head is positioned over its respective track on the drum.

We have generally accepted a memory drum as the most suitable choice of memory devices. It will rotate at as precisely constant a speed as possible. We will consider its application in greater detail after relating it to our original example where two 3/2 frequency conversion steps of a sequence were given.

The output frequency of step one appearing at unit two, is now routed to the memory. After an adequate lapse of record time has been provided, at least equivalent to the time needed for a full operation of the frequency converter, the input signal frequency of the first step is removed. The recorded frequency is then played back for the second operation as an input to drive the third unit and in turn produces the output appearing at the second unit. This in turn is recorded on another track before removing the input. The routing of the music progression would be programmed by an operator either from a keyboard related in intervals, or, as an improved facility, from a punched tape or equivalent.

The input frequency for a given operation would have

to be of sufficient duration to allow the motor speed to synchronize and then to relay the output of the alternator back to a record head on another track. The recording is then stored as a step in the final performance sequence and is also ready for playback in the next operation of the frequency-to-frequency assembly sequence.

Returning to our consideration of the memory drum, a problem may arise particularly when the output frequency of the alternator is being recorded on a given track. Since the memory drum diameter is fixed and the speed of rotation is constant, the time from the instant a frequency is recorded on a track to the instant it will return to that point must be taken into consideration for two reasons. The frequency at the beginning and at the end of the recording track may not be in the same phase relationship to one another. Secondly, the record head must be told when to stop recording so that pointless re-recording does not result. To provide for these two needs a longitudinal reference line must be indicated along the surface of the drum, that is, parallel to the drum's axis of rotation, to serve as the reference point where it passes through a given track for the start and stop sequence of the record and also for the playback operation.

However, another design problem remains since the total time required for playback from the memory in a given operation is greater than that available in a single revolution of the drum. The operation involves first synchronizing a given motor unit to the new input frequency and then making a recording from its related alternator output. The duration of the recording requires a full revolution of the memory drum. The memory drum, in turning more than one complete revolution, will cross the start/ stop reference point at least once. This means that the possible out of phase relationship at the reference point may carry over this slight error by introducing it to the motor unit it is driving. After which, the error may be transmitted to the output alternator that will record the next successive track in the musical sequence with a potential phase shift error introduced onto it.

This phase shift error will be less than one cycle per second divided by the product of the playback (or frequency converter input) frequency in c.p.s., times the number of seconds for one revolution of the memory drum. The possible frequency shift error should have negligible effect on the frequency converter motor unit. However, it would seem preferable to isolate the start/ stop reference point as much as possible. This can be done in conjunction with the prior suggestion of making a reference point common to all tracks, that is, a fixed common point on the drum's rotation.

Therefore, the playback track would drive the given motor unit into frequency synchronization. When the motor unit is synchronized, the respective alternator would start recording the new note frequency on the successive track at the given instant when the memory drum's reference point crossed the record head. This will synchronize the record head and the playback head of a given frequency conversion operation in the sense that both will cross the reference point simultaneously. This will eliminate introduction of potential cumulative errors.

Although it is preferable that the playback heads be staggered around the periphery of the drum for the final playback, for the frequency conversion record/playback sequence, a record/playback head that can sweep the memory drum's outer periphery from track to track paralleling the memory drum's axis of rotation is favorable.

This would suggest that the frequency conversion operation utilize a playback head situated over one track, with a record head situated over the next adjacent track. The pair of movable heads would be sequentially located by a micrometer drive. The drive would be in stepped motion going from track pair in operation to overlapping track pair in operation, i.e., playback track 1, record track 2; P.2, R.3; P.3, R.4; etc.

The object is to try to keep the latent trouble-maker, the potential phase shift at the reference point, out of the frequency conversion and final playback recording operations. While the frequency conversion operation is one requiring a *fixed* sequential routine, the playback operation conflicts in processing requirements by being a *random* operation of both the sequence and duration of the horizontal musical movement of the composition. This difference will be elaborated on in the following consideration of how to remove the recorded intelligence from the memory drum in final playback.

There is an additional situation that should be covered insofar as it may provide a detectable source of unintended sound. During the final playback operation, the sequence of note frequencies will be taken from their locations on different memory drum tracks at a rate of speed equivalent to the final performance. If a given segment of the musical sequence should have three notes on it, one a second in duration, the next two each a half second in duration, then this is exactly the duration that their respective tracks will be in operation. That is, the first head to be used for playback will be switched into its respective waveshaping circuit for exactly one second. The next two heads will be switched in for a half second apiece. Therefore, wherever the playback head in use at a given instant is located on its track, it will be switched into its waveshaper. The problem is that there is a chance that it will cross over the instant start/stop reference point from the original memory drum recording process. The result will be that in the course of the frequency's duration, should there be an out of phase relation at the instant start/stop reference point it may show up. This error can travel right through the associated circuits and may be detectable in the finished composition at perform-

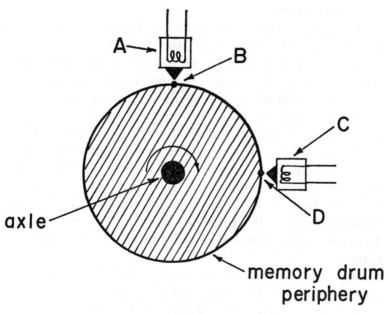

Fig. 19

ance time. To solve this potential problem, we will go directly to the source and eliminate it.

It is assumed at this point that there is a playback head

for each track. This will later be shown as the preferable design. Further, as in computer practice, the heads are staggered over the periphery of the drum in order to have room for placement. If we were to consider a given instant during the continuous revolution of the memory drum, we would see that since the heads are all located in different positions on the memory drum periphery, if two or more were in simultaneous playback, they would be receiving a signal in turn from different lateral locations of the drum's periphery. We will now use this fact to our advantage.

Our start/stop reference point on the memory drum is a line parallel to the axis and therefore it will occur for all tracks at a particular instant. However, our heads are staggered, so a given point of drum rotation could never correspond with all of the heads simultaneously. If our start/stop reference line were under a head at a given instant of time, all other heads, by being located at other positions of the drum's periphery, would not be crossing the same reference point. We can see this in Figure 19. The start/stop reference line B is under head A and therefore head C, by being in a different position, is not simultaneously at the reference line. Points B and D are locations on two adjacent recording tracks of the memory drum.

The memory drum is now in the final playback mode and is providing frequency information, previously recorded on it, to the wave shapers and other circuitry at final performance speed. Again referring to Figure 19, if head A is in the process of playing back a note frequency and by chance happens to cross point B during the required playback duration, any possible by-products of the start/stop crossover point will be included in the final recording. However, head C, which should be the next head in the playback sequence is nowhere near the start/stop reference point B, and therefore it can be used as an alternative track. When there is a need for an alternate selection of track, this must be anticipated in advance of the problem. The way we will arrive at suitable means will be shown shortly. First, an additional operation must be contemplated.

Since the memory drum's rate of rotation is constant throughout all operations, it is logical to assume that if we were to fix an arbitrary point on it and designate this in a given instant as the starting point of the final playback of the completely programmed memory drum, so many revolutions later at another given instant the composition would be completed. This time parcel for final performance can be further divided into segments, each of which constitutes a period when one particular note, or simultaneous notes, in the overall sequence will occur. Our additional operation involves information that is recorded on the punched tape or equivalent master program. The final playback time sequence will now be referred to twice in the overall process of composition programming. Previously, it was used to control the recovery of note frequencies from the memory drum for shaping into the final performance at normal performance speed. This would occur after the operation utilizing the frequency converter to program the memory drum. Now, this timing track will also be used before the frequency converter operation is started.

A point of reference representing the instant the composition will start at its normal performance speed is located on the memory drum. This point would most likely be the location of the first of the staggered playback heads at the start/stop reference point. The memory drum is started and after it is operating at its correct constant speed, a circuit triggers the punched tape into operation. A circuit will then compare the sequence of the heads for the final playback performance sequence with the time duration of each note as taken from the musical timing portion of the punched tape. If the start/stop point is to be crossed in the course of the note frequency as recorded on its correct track, the compare circuit will be activated. The compare circuit shifts a

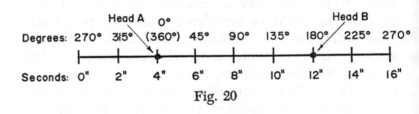

Fig. 20

preset circuit (to be later referred to in the general frequency conversion operation) that will eliminate the potential trouble track from the sequence. The compare circuit skips the trouble track and jumps the problem note frequency to the next track in the sequence. This alternate track will not have the start/stop point in the same proximity as did the previous one, and therefore the likelihood of the problem reoccurring on the alternate tract is almost impossible.

There is no reason why the compare circuit can not be

scanning more than two successive circuits at a time for suitable placement of a note frequency on a track. That is, it can scan the next successive track in the horizontal musical note sequence and one or more successive alternate tracks. Such steps are most likely unnecessary but require little additional circuitry. Also, the probabilities of two problem tracks in a sequence for a given time lapse of a note frequency can be readily controlled by the variable design factors of memory drum diameter and its speed of rotation.

An example of the foregoing might be in order. In Figure 20 a complete rotation of the memory drum is shown, where the reference start/stop point is given to be at the zero degree point. We are in the *first* programming operation, where the timing of the notes of the composition is being compared to the portions of the tracks on which they would normally occur unless altered by the compare circuit. Head A has been programmed to receive a four second note frequency that will occur from 315 degrees to 45 degrees (with reference to the zero position of the start/stop point). As has been shown in Figure 20, one complete rotation will require 16 seconds. Midway in the compare check, the reference start/stop point appears. The compare circuit, which would most likely be composed of electronic switching, switches head A out of the sequence and replaces it with head B within micro-seconds.

At the same time, a preset is switched to remove the trouble track from the frequency conversion operation. Therefore, the stepped micrometer drive of the frequency converter will be programmed to skip one track in the recording sequence. This can be done by a temporary mechanical shifting of *only* the record head of the record/playback head pair to a position two tracks away from the playback head. For example, if the playback head were on track 10, then the record head would be on track 12. This special head position would automatically be tripped back to the normal record/playback adjacent track relation on the next succeeding operation. Therefore, the track sequence and related notation would be: 9-10; 10-(11), the problem track is noted and spacing of record/playback head is extended to span three tracks, as 10-12; the stepped micrometer drive skips the trouble track, (11), and the record/playback head spacing is returned to its normal span of two tracks, as 12-13.

Continuing on, heads A and B, at the given instant in Figure 20, are at the zero degree and the 180 degree point respectively. Therefore, the four second note will appear on the track under head B in the arc circumscribed between 135 and 225 degrees. The start/stop reference point has been eliminated.

In addition, the musical note frequency sequence will pass through harmonic assembly. That is, the sequence can be distinguished as taking place both horizontally and vertically. The horizontal melodic progression can swing to a vertical harmonic progression, then return to the equilateral pivot note frequency to continue horizontally. This would appear as a memory drum track sequence, such as: 3-4, horizontal; 4-5, vertical; 4-6, vertical; 4-7, horizontal. The track numbers refer to Figure 26. Therefore, additional control of the track span mechanism of the movable frequency converter playback/record head pair is implied. Wherever directional change occurred in the musical note sequence with a note frequency to be *carried*, the information for the necessary control would be placed on the punched tape.

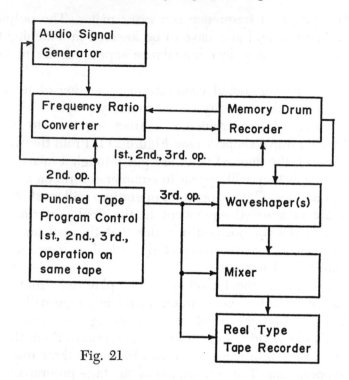

Fig. 21

The frequency converter head positioning mechanism used for the control of track spanning and associated micrometer drive can be replaced by a single row of combination record/playback heads spaced as closely together as possible. They are placed in a row along the outer periphery of the memory drum, parallel to the axis of rotation. While this alternative should simplify the mechanism required, it will reduce the possible number of tracks per given width.

The playback signal from the memory drum would most likely be a low level, high impedance signal. This signal would require amplification to a power level for driving a given frequency converter motor. The output impedance would also have to be lowered in matching it to the most likely low impedance inputs of the motor units.

After the completed composition or section of it, depending on the capacity of the memory playback drum, the frequencies are routed in correct sequence to their respective wave shapers (see Figure 21). From the wave shaper(s), the musical notes are put through a mixer and recorded as they will appear in completed form on a reel-type recorder. This sequence of operations will most likely operate at a speed equivalent to the final performance speed of the composition in order that the reel tape recorder can record the completed musical sequence in an uninterrupted manner.

To facilitate the handling of the playback from the memory drum to reel recorder, a punched tape will indicate the composer's intentions. The coding of the punched tape can be divided into two distinct sections. From these two sections, information is drawn to control three respective operations. The first section of the tape programs the frequency converter sequence. The second section controls the timing of the final playback along with other pertinent information for completing the composition. The data from the first and second section progress at an equal rate along the length of the punched tape.

The punched tape will be read three separate times. That is, there will be a separate pass for each operation. The first operation will draw timing information from the second section of the punched tape to feed the compare circuit. The compare circuit searches the memory drum for a random start/stop reference point crossover during a note frequency recording and adjusts a preset pattern accordingly.

The second operation will draw information from the first section of the punched tape to control the compatible operation of the frequency ratio converter and the memory drum.

The third operation is controlled by the second section of the punched tape. In the third operation of the punched tape the memory drum is played back in the required sequence and individual note frequencies are fed into their specified wave shapers. The punched tape also controls the mixer to balance the amplitude outputs when a plurality of different wave shape categories are occurring simultaneously. During this third operation, the tape recorder would be running at a continuous steady speed equivalent to what the playback speed of the final tape would be in performance, such as the standard tape speeds of 7½ or 15 i.p.s.

Before attending to the details of each unit, an example

Moderato

Fig. 22

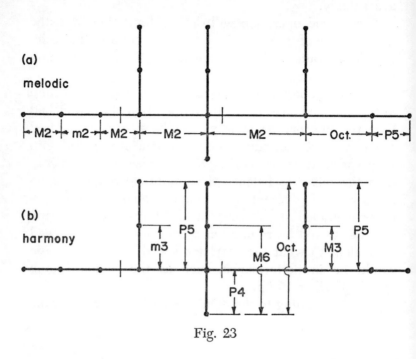

Fig. 23

of composition would be in order to recognize what some of the possible unit requirements will be. At the same time we have the opportunity to examine a possible form of notation for the composer and see how it might develop from our elementary forms of notation previously used in illustration. For our sampler a conservative passage is intentionally used, one easily recognized by its conventional harmony and modulation. It is shown in Figure 22 as it would appear scored in the treble clef.

In Figure 23 the musical phrase is transferred to notation as would apply for a single instrument capable of producing harmony. That is, the completed phrase could be given the timbre, or waveshape, of a voice of the pipe organ. In Figures 23(a) and (b) the notation is shown as it has appeared in previous examples of step-by-step movement: in (a) the melodic step-by-step motion is noted and in (b) the harmonic bottom-up assembly is

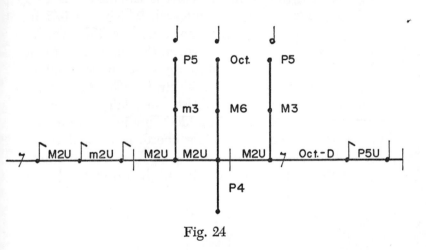

Fig. 24

shown. The intervals in both (a) and (b) occur together but are shown separately for ease of illustration.

In Figure 24 the dots, or points where notes are designated to appear, are replaced by conventional notation in order to give some reference to the elapsed time for each consecutive note in relation to a given fixed meter, or rhythm. In the case of a chord it appears designated above the dots. Also added to the interval notation is a suffix that is either the letter U, for up in frequency, or D, for down in frequency. This identifies the direction in which the interval is moving, ascending or descending in frequency from the previous note step. The first symbol on the melodic step progression line is an eighth note rest as used in conventional practice.

In Figure 24 the relation of notes in the harmony are assumed to be assembled from the base, or lowest note, of the simultaneous note group. Therefore, with reference to Figure 23, the F minor chord is assembled as F up to A flat, or a minor third interval, which is a 6/5 frequency increase, and then F up to C, or a fifth, which is a 3/2 frequency increase. However, the notation of Figure 23 is abbreviated since its purpose is now understood and it is no longer necessary to illustrate it in the previous manner of a chord assembly. Also, as explained previously, the G major chord with the added note in octave relation that is succeeding the F minor chord, is assembled from the melodic note G down to the D, then from the lower D, or bottom note, up to the B and then from the bottom D up to the top D.

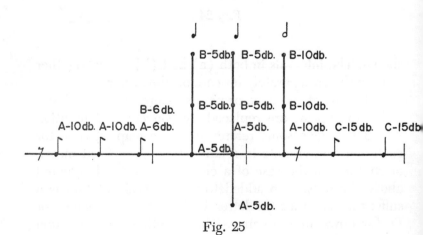

Fig. 25

Up to this point the notation as provided in Figure 24, takes into consideration the use of only one instrument, or one type of timbre. In Figure 25, which is developed from Figure 24, additional notation is included for the scoring of a hypothetical group of instruments, each of which would have a specified timbre or range of timbre. For simplicity, only the instruments given as A, B, C and their respective amplitude levels are shown. Figures 25 and 24 may be combined into one.

In previously suggesting that there might be a range of timbre for an instrument, let us use the clarinet for one example. If we are choosing to produce as close a replica of timbre to the clarinet as is possible and are not trying to isolate our new instrumental sounds from conventional instrument sounds, then we must first note timbre deviations over the range of conventional instruments. For example, the clarinet may be divided into at least two separate registers. The lower is commonly termed the *chalumeau* and the upper is the *clarion*. The clarinet is for our consideration actually more than a single instrument with regard to timbre. Therefore, a full understanding of its range must be anticipated and included within the programming of the composition in order to obtain exact duplication. Sounds not meant to duplicate conventional instruments can be regarded as being of one harmonic type, or timbre. Except when a conventional instrument is being duplicated, it is irrelevant whether two or more different sounds represent the different registers of one conventional instrument or are being considered as independent original sounds. This is where we will stop in our illustration since we are getting into the area of actual creation in which the distinctions are the composer's domain.

The amplitudes of our three distinct timbres in Figure 25 will need to be considered. This would concern their balance with regard to one another and their overall level. This is comparable to prior dynamic notation in conventional orchestration; the one difference is that previously, in order to achieve an orchestral balance, it was assumed that, for example, the potential amplitude of one trumpet would take a number of flutes to equal it. Until now, the overall choice in the balancing of simultaneous instruments was a delicate matter of experience. Now both the instrument and its amplitude are controlled without any limitations other than those imposed by the composer. If we previously wanted one trumpet to be of equal amplitude to a flute, we needed for our balancing means a group of flutes in unison with each trumpet. Now we need only raise the amplitude level of our flute.

The foregoing mention of balance of combined different timbres is already in the practiced and growing domain of electronic music, or music whose sounds originate from electronic sources. A number of composers who started their careers by creating for conventional instruments have already transferred their orchestral experience over to purely electronic techniques; such experience would be applicable here. Because such knowledge is already existent there is no point in further digressing into the means and balance of the wave shapes that will be on our final recording.

Having derived Figures 24 and 25 from our musical example of Figure 22, we can now see more clearly what some of the general requirements for our processing equipment will be. With this information and by using Figures 24 and 25 for examples, we can return to our

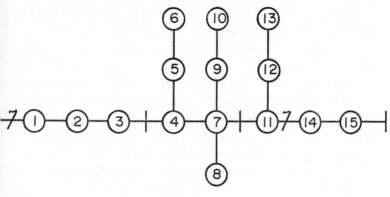

Fig. 26

consideration of the individual block units and what they might be required to do.

The second operation of the punched-tape program sequence for controlling the frequency ratio converter in conjunction with the memory drum is taken to be completed. The drum has recorded a sequence of frequencies in step-by-step movement. It will start the sequence from the first or fundamental frequency selected and controlled by the punched tape and produced by an audio generator. From this point forward the ratio relations of the frequencies are determined in step-by-step fashion by the equipment used in the second tape operation. The sequence for the musical example given in Figure 22 is shown in Figure 26. Each numbered step in Figure 26 would be a frequency that is recorded in the same sequence onto adjacent tracks of the memory drum. Figure

26 illustrates the step-by-step assembly process: we see that the frequency note 7 extends as an interval from frequency note 4; also 11 extends in interval relationship from 7, and 14 extends as an interval from 11.

With regard to the melodic-harmonic sequence interrelation shown in Figure 26, we can observe that the second operation of our tape program will have two additional intelligence steps to perform while extracting the track frequencies from the memory drum during playback. It will have to know when to alter the continuous movement from adjacent track to track, that is, how and when to jump about from track to track during memory drum playback. It will also have to know where and what tracks are composed of a simultaneously occurring group, or musical harmony. It will have to know if one of the tracks composited into a simultaneous event involving several tracks is to be held over while others are released. Conversely, it will have to know if a group of tracks are being held simultaneously, as for producing a chord, while a sequence, such as a melody, is occurring superimposed.

For examples of the foregoing three general situations we can generally refer to Figure 26. At note frequency 4, 5 and 6 are occurring simultaneously as a harmony. From note frequency 4 the melodic progression jumps to track 7. The simultaneous note frequencies 11, 12, 13 are being held while the melodic sequence 14-15, starting a half-beat later, is being superimposed on the chord.

It should be noted that this is understood to be by no means an exhaustive analysis of the three basic possibilities for utilizing track sequence on the memory drum during playback. However, the foregoing also serves to illustrate that more than one final playback head is necessary at the memory drum. The number of heads required are at least as great as the number of simultaneous notes in a given composition.

Moving on from the memory drum, it is in the wave shaper that each note frequency must be identified by the punched tape as to its timbre where there is more than one type of timbre used within the composition. In Figure 25, three different timbres—A, B, and C—were present. Sometimes they occurred together: A and B occur simultaneously at the point designated 3 in Figure 26. Also, at simultaneous group 4-5-6, the A has a separate timbre from the B's and so they must be routed to two different wave shapers.

Regarding wave shaping, comparison can be made to general forms of electronic music and need not be dwelled upon at length here. Briefly, the sine wave can be handled in a formant circuit, which is most likely to be the selected procedure here. By formant we mean that the sine wave, which up to now was distinguished from the finished note product by being termed a note frequency, enters the wave shaper as an input and exits as any complex wave intended; the input sine wave frequency is the fundamental and harmonics are *shaped* into, or added to, it. To be a little more explicit, the sine wave triggers a complex wave generator. The harmonics of the resulting complex wave are in turn filtered, that is, subtracted or distorted, to get the required results.

The general alternative frequently found within electronic music procedure is that of synthesis. Synthesis is building up a musical note by an additive process basically following the Fourier analysis that any complex wave is made up of sine waves in simple ratio relation to one another. It is important to know the waves' phase relation to one another as their interaction can cause modulation or cancelling of their components. There are also other parts of the complex wave that must be considered in synthesis and that can not be directly broken down into simple waves; elements of noise are such parts. Therefore synthesis essentially produces the musical note by repeating a pass over a given tape length, each time adding another sine wave, or element of the complex whole, while control is provided over the various phase relations and amplitudes.

One of the problems in synthesis is the accumulation of noise on the tape by making repeated passes. Another problem is the matter of successfully interpreting all the basic elements composing the complex wave, or timbre, of a musical note. A third is retaining the correct phasing of all the ingredients. Synthesis, due to the potential complexity that can arise, is generally looked at today as being more successful in theory than practice, and so we can dismiss it from our use.

Having observed what need be controlled at the wave shaper by the punched tape, we can move on to the mixer. The mixer controls the amplitude of the composition's wave shapes if there are several occurring at once. That is, their balance with one another and their over-all amplitude is controlled. The information required by the mixer is provided by the punched tape. The minimum number of channels in the mixer would be equal to the greatest number of simultaneous wave shapes occurring at any one point in the composition. Each wave shaper would follow directly into a channel of the mixer. In the mixer the various wave-shaper channels are blended into one output. This makes the output of the mixer at any given instant the result of all the preceding intelligence provided by the punched tape program. This single, complex, wave envelope is then recorded onto a tape passing by the record head in a steady and continuous motion. Upon playback the tape will provide the sound of the completed composition, or sequence, depending on the programming capacity of the memory drum unit.

As noted in Figure 25, the various amplitudes of each individual note are specified in decibels from a given known reference point of sound intensity, or zero level. In Figure 25 0 db. is considered the threshold of audibility. All amplitude information would be provided by the punched-tape program for the mixer during actual programming. Referring to Figures 25 and 26, the composite simultaneous levels of points 11, 12, and 13 are each 10 db. Since the levels of 11, 12, and 13 are logarithmic relations, they may be added to show that the output from the mixer at that given instant will be 30 db. above the selected zero reference level.

Returning to our consideration of the second operation of programming, the frequency ratio converter was noted to require synchronizing time so that the input motor could be brought into step with the playback frequency from the memory drum. This time delay, along with the possible desire to design the frequency handling along completely electronic lines, would indicate the substitution of a digital or analog type frequency converter.

A frequency converter can take many forms. Let us first review our design limitations. In changing electronically from frequency to frequency in a ratio, both a multiplication and a division unit are required. The units must operate from a variable input. Also, they must produce a linear output, that is, an output evenly spaced in time. Probably pulses would be the processing medium that would trigger the wave shaper units on memory drum playback.

The frequency divider that can accept and process a variable input frequency can be a circuit frequently seen as a computer-type counter, which is an aperiodic multivibrator, or flip-flop. Any series of divisions can be provided by correctly ordering the necessary number of flip-flops with the required feedback circuits. The output would be a series of periodic pulses with a time lag between each of them that would account for the input pulses that had been suppressed in order to provide a reduced pulse output. By allowing one pulse to pass from input to output and then suppressing one or more pulses after it and then repeating this cycle, a division is in effect taking place.

Frequency multiplication is normally accomplished in applications where the input frequency is fixed and the output is a harmonic multiple of the input frequency. The extracted harmonic is the most resonant one going through the tuned output circuit. The tuned output is set to peak the desired harmonic while attenuating all others present. This would not be suitable for a variable frequency input. Therefore, one possible circuit is suggested where the input pulse would be converted into a sawtooth pulse. The sawtooth provides a rising linear voltage ramp for each cycle of input frequency pulsing. The ramp of the sawtooth is kept as linear as possible. Associated with the sawtooth voltage is a sequence of triggering circuits each set to go off at a different fixed voltage level of the sawtooth ramp. These associated circuits will be in a sequence and will provide a pulse that is their part of the output pulse. Fine adjustment for the exact firing level could be provided for each associated output pulse circuit. Their interrelation could be aligned on an oscilloscope, synchronized to one cycle of the sawtooth input wave, so as to adjust them to be exactly symmetrical with each other in time.

If we want a frequency multiplication of three, then three associated trigger circuits are used in conjunction with the sawtooth firing voltage circuit. Three separate and differently phased output pulses, which are combined into a continuous sequence at the output, will result. For each symmetrical input pulse there are now three symmetrical output pulses.

An alternate frequency multiplier works with sine waves, and although quite simple, it is limited in usefulness since it can multiply only by powers of two. A sine wave input is put through a full wave rectifier that forms the output. The full wave rectifier has a ripple frequency that is twice that of the input frequency. If the output of our full wave rectifier is transformer coupled, or treated by means giving similar results, we now have converted our output sine pulse back into a sine wave. We reapply the wave to another full wave rectifier to quadruple the input-to-output frequency relation.

To illustrate the foregoing means of conversion, we can change a frequency by 5/3 or a major sixth up. Whether we multiply the input pulse from the memory drum first by five and then divide by three, or vice versa, is incidental. If we choose first to multiply by five, we will convert the input pulse to a sawtooth wave that will fire a sequence of five associated circuits, each one set to be triggered off at a successively higher voltage. The resulting pulses are combined at the output and are then fed to the necessary flip-flops. In this case we are dividing by three and therefore require two flip-flops and one feedback circuit. Our output is now a pulse rate 5/3 the input pulse rate that is recorded on the next track of the memory drum.

Technique generally following analog practice appears to run into problems since the input frequency must be converted to a voltage, stored, and then converted back to a frequency. The ratio of voltage change is thus related to the required rate of frequency change. Keeping the frequency ratio conversion exact and needing a steady voltage as the memory playback are two problems confronting this area of investigation.

Before concluding the outline of a preliminary design, an alternate to the tape drum memory should be briefly covered. If the memory drum crossover should prove in prototype testing to be objectionable, the following may be chosen as an alternate means of avoiding reproduction of the crossover phase shift.

In this alternate method, the memory no longer is a magnetic tape loop. It is now an open ended ribbon. Two special purpose tape decks would be required. One would be a magnetic tape transport adapted from a digital computer using vacuum tape buffers. The other deck would be similar to a regular audio deck but with special layout of the components.

A magnetic tape unit from a digital computer possesses certain features desirable here. For example, it is important to be able to accelerate the tape from a standstill to a very steady operating speed in a few milliseconds. Since the tape passes the head at 30 or more inches per second, depending on the unit, accelerating to this speed from at rest in a few milliseconds without any lurching (wow or flutter) is a significant feature. This feature is primarily due to use of the vacuum buffers. There are two

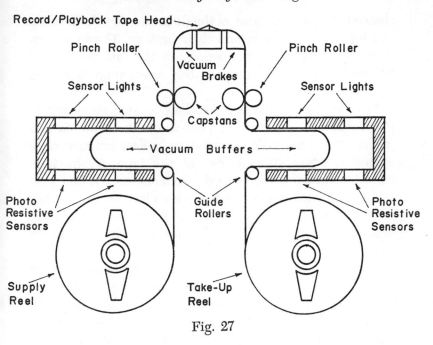

Fig. 27

buffers located between each of the reels and the head, as shown in Figure 27. The buffer is a deep chamber of rectangular shape with one end closed. The width of the buffer is a very close fit to the width of the tape, allowing the tape to fit snugly in the chamber. As the tape comes off either reel, it forms a loop into the open end of the buffer before passing around the head. A vacuum is created within the closed end of the buffer to retain the tape loop. The vacuum is formed by a vacuum pump connected by a tube to the closed end of the buffer. Near the entrance and near the closed end of the buffer are two sensors that can detect whether the loop is entering too far or insufficiently far into the buffer box. The sensor can be a vacuum switch or a photoresistive, light-sensing element. The sensors control the direction that the take-up and feed motors run in and their braking. The tape loop is stabilized between the sensors in the buffer chamber although a constant compensating fluctuation is taking place.

To either side of the tape head are located pinch rollers that engage the tape and drive it across the head. As this is done the sensors in the vacuum buffers automatically control the take-up and feed of the tape. Again, it is important to emphasize our dependency on the fact that a few milliseconds after the pinch rollers have been engaged the tape is passing at a very steady rate across the head. Only a tape deck of this level of control can be considered.

The second tape deck generally follows standard audio design and operates at a continuous speed. The tape feed is assumed to be as steady as is commercially obtainable. What makes the unit non-standard is the large separation of the record and playback heads that are meant to operate simultaneously. The tape covering the delay between the heads could be made more compact by travelling through a zig-zag idler configuration, or a specially elongated tape transport could be utilized. The frequency conversion process will be handled first by this tape deck.

In choosing from previously discussed means of frequency-to-frequency conversion it is assumed that the processing will be totally electronic. This is necessary to meet the required speed of operation. The distance between the record and playback heads allows for a time delay that is of greater duration than the longest single frequency in the composition.

Processing a composition is started by a single frequency

being recorded from the predetermined duration. It is a randomly chosen frequency taken from a suitable generator in accordance with previous explanation. Looking in the direction of tape flow, the record head precedes the playback head. The recorded frequency is completed at the point where it extends on tape, the gap between the two heads. Just after completion of the recording, the recorded material starts crossing the playback head. The playback head in turn feeds the original frequency through electronic frequency-conversion processing so it exists as a new frequency, changed by a simple ratio. This is immediately transferred to the record head so that the new frequency is being recorded simultaneously and is in step with the frequency now in the playback mode.

Between the recorded frequencies there is a short delay so that a reader, which is in step with the tape deck and which has just timed the duration of the recorded frequency, can now connect the playback frequency to the desired electronic frequency changer. The ratio of frequency change has already been programmed onto the paper control tape. The recorded frequency now at the playback head is switched by a reader to a processor where the frequency is converted by simple ratio and then transferred back to the record head, where it appears as the next frequency in the sequence. The process is basically similar to that used for the alternate memory drum.

Each frequency on the completed magnetic tape will have the same fixed duration. However, for the completed performance, the sequence of frequencies will undoubtedly vary in duration. Also, their harmonic interplay must now be routed. The computer transport, our second tape deck, is now required.

The first tape deck is again set in steady operation. On it is the recorded magnetic tape. Only its playback head is now being used, which is coupled to the record head of the second deck. Where the first deck required only a single track, the second would favor as many tracks as are available. Nine tracks would be available if a standard computer head were applied.

A control unit is required for operation. It will operate continuously and must be in step with the frequency recordings on the first tape deck. Therefore, if a paper-tape reader is again utilized, having on it the program of this second operation, it must be positively connected to the magnetic tape drive. The control unit will start the second deck by activating the pinch roller solenoid when a signal is received just prior to the given recorded frequency. Furthermore, the control unit will stop the second deck while allowing the first deck to continue on at a steady rate, all in accordance with the timing specified on the paper tape for each given frequency.

The control unit will also route a given frequency in order to locate which track it will be placed on in the case of simultaneous and overlapping frequencies. Where intended, it will return the second tape deck to the starting position of a frequency just recorded so that other frequencies may be recorded beside it.

As a possible alternative, the first tape deck could also have rapid start/stop capability like the second unit. Both would be simultaneously triggered at the start of each frequency. Therefore, the first deck could be left idle while any required control signals set up the second deck for the next frequency.

Wave shaping and dynamics could be programmed in

when transferring from the first to the second tape deck by synchronized control units working with the already mentioned transfer control unit for timing and track location. Again, the details would generally follow procedure already outlined for the alternate memory drum.

6
The Author's Corner

Perhaps the worst misunderstanding in contemporary music, apparently encouraged by the many unrelenting ties with tradition, is that music generated by electronic instruments is inferior to a live performance of non-electronic instruments. Whether it be a concert that is being critically reexamined in a music class or the casual background music in a supermarket, people of varied musical backgrounds are unifiedly accepting electronic reproductions of live non-electronic music and rarely or never hearing their live-performance counterparts. This deception of electronically reproducing sound to make it appear that a live performance is taking place operates with varying success, depending on the efforts expended and on the quality of the reproducing equipment. But almost regardless of the varying quality of reproduction, the listener is still unquestionably accepting what he hears as music. Electronic reproduction is inexpensive, easily available, private and therefore quite realistic. Why then is there so much public opposition to electronically generated music by the same people who accept electronically reproduced non-electronic live music?

The same technology that deftly reproduces a live performance can generate the equivalent "artificial sound." Tests such as those conducted with the R.C.A. Synthesizer have shown that a large body (70%) of the listeners cannot discern the live instruments from their synthesized equivalents. And the Synthesizer, like everything mortally contrived, can be improved upon to deceive even more people. Electronic music generation has no real limits, whether in duplicating conventional sounds or in providing novel but nonetheless worthwhile sounds.

It would seem that to electronically duplicate conventional sounds does not give the musically involved public much reason to become excited. The technical feat is very worthwhile. But the listener is not particularly appreciative of technical feats unless they provide something not previously possible. Certainly, performers and advocates of conventional instruments will not concede any merit to electronic music generation if their position is being challenged by an upstart that has to struggle through very sophisticated routes just to approach equaling them.

Possibly frightened by the "yardstick" of conventional music, composers involved in electronic music generation have frequently gone to extremes to show the public that their product is both novel and worthwhile. Unfortunately, the welding of an expert composer and an electronic technician into one man is not a common event. An experienced composer for conventional instruments all too often seems lost in attempting to translate his musical desires into an electronic music composition. The electrotechnical background required can be overwhelming. In turn, the designer of electronic music instruments is dazzling the composer with the potential of electronic music generation but he cannot provide the formula of how to metamorphose control-panel elements of music into art. The result is all too often the sounds of experimentation, insufficiently edited! (Sometimes judicial editing can be observed where no other material was available to replace that removed. An example exists in music composed of several minutes of inspired silence.)

Such problems only confirm to the conservative composer, performer, and listener that electronic music generation is in the extreme area of the state of the art. It is adjudged a distinctly inferior and therefore disassociated environment from that of live music reproduction.

Possibly an additional direction is now needed. A middle ground, satisfactory to the majority, would provide a course linked directly with our two-thousand-year-old heritage. The greatest emphasis will again be properly placed by the composer on pleasing his music audience rather than pleasing other composers by his literary excellence. There is too much emphasis on the level of technical involvement and the status climb is losing sight of the humble purist's pursuit. With all of our technological advances, especially in this century, there are too many people who feel lost in the complexity and are expressing their inability to master the new media by building empty fronts of intellectualism; they are tacking up false standards to disguise their own minor offerings to society. The technological materials that we have to work with today are reservoirs of tremendous wealth and betterment, awaiting only the discovery of applications. But the evolution of stages in art must be logical and not disjointed. The radical opportunists who leap after novelty for their own name's sake will eventually become just forgotten history. This façade will crumble as electronic applications mature in music.

Total Consonance is meant to be a conservative next step for music. It is certainly a unique enough approach to give electronic sound generation a purpose that no fixed pitch instrument orientation can equal. Its success will require the acceptance by good composers, conservatives who never jump at a new idea. People who rush for extreme novelty are not the people to shoulder the massive labor of improving what are already our guidelines for excellence. Excellence has already been extracted by numerous geniuses from the duodecimal equal temperament system that is foremost today. Total Consonance has a great deal of work ahead of it before it will be publicly secure. But, in its favor, the improvements are in keeping with tried and proven century-old objectives. I believe it still takes more to make music today than merely to call something "musical." If these closing remarks spark a fire—burn, fire, burn!

References

Gottlieb, Irving M. *Frequency Changers*. Indianapolis: Howard W. Sams, 1965.

Helmholtz, Hermann L. F. *On the Sensations of Tone*. New York: Dover, 1954.

Jeans, Sir James. *Science and Music*. New York: Cambridge University Press, 1961.

Lytel, Allan. *Fundamentals of Data Processing*. Indianapolis: Howard W. Sams, 1964.

Olson, Harry F. *Music, Physics and Engineering*. New York: Dover, 1952.

Technical Education and Management, Inc. *Computer Basics*. 5 vols. Indianapolis: Howard W. Sams, 1961.

Wood, Alexander. *Acoustics*. New York: Dover, 1966.

——. *The Physics of Music*. New York: Dover, 1961.

Yasser, Joseph. *A Theory of Evolving Tonality*. New York: American Library of Musicology, 1932.

Index

107